IES
D0475641

Here Comes the Sun

Also by Steve Jones

The Language of the Genes
In the Blood
Almost Like a Whale
Y: The Descent of Men
The Single Helix
Coral: A Pessimist in Paradise
Darwin's Island
The Serpent's Promise
No Need for Geniuses

Here Comes the Sun

How it feeds us, kills us, heals us and makes us what we are

STEVE JONES

Little, Brown

To the memory of George Harrison, who brought sunlight into so many lives

LITTLE, BROWN

First published in Great Britain in 2019 by Little, Brown

1 3 5 7 9 10 8 6 4 2

George Harrison quote from his book *I, Me, Mine*, Genesis Publications, London, 1980.

A CIP catalogue record for this book is available from the British Library.

Hardback ISBN 978-1-4087-1131-6
Trade paperback ISBN 978-1-4087-1130-9

Typeset in Bembo by M Rules
Printed and bound in Great Britain by Clays Ltd, Elcograf S.p.A.

Papers used by Little, Brown are from well-managed forests and other responsible sources.

Little, Brown
An imprint of
Little, Brown Book Group
Carmelite House
50 Victoria Embankment
London EC4Y 0DZ

An Hachette UK Company
www.hachette.co.uk

www.littlebrown.co.uk

And seeing the snail, which everywhere doth roam,
Carrying his own house still, still is at home,
Follow – for he is easy paced – this snail,
Be thine own Palace, or the world's thy gaol.

JOHN DONNE,
'Verse Letter to Sir Henry Wotton' (1635)

CONTENTS

PREFACE

THE CRY OF THE EAST WIND

There are two seasons in Scotland: June and winter.

Billy Connolly

In 1938 Joseph Goebbels saw some pictures of Edinburgh. He was impressed, and noted that 'It will make a delightful summer capital when we invade Britain.' The photographer must have been lucky with the weather, for a native of the city was less positive:

> Edinburgh pays cruelly for her high seat in one of the vilest climates under heaven. She is liable to be beaten upon by all the winds that blow, to be drenched with rain, to be buried in cold sea fogs out of the east, and powdered with the snow as it comes flying southward from the Highland hills. The weather is raw and boisterous in winter, shifty and ungenial in summer, and a downright meteorological purgatory in the spring. The delicate die early and I, as a survivor, among bleak winds and plumping rain, have been sometimes tempted to envy them their

> fate . . . Happy the passengers who shake off the dust of Edinburgh, and have heard for the last time the cry of the east wind among her chimney-tops!

Thus Robert Louis Stevenson, and he lived up to his words, for he abandoned the Scottish capital for California, only to expire in his mid-forties under the blue skies of Samoa.

I arrived in the city as a callow undergraduate in 1962, in time for the coldest winter of the twentieth century, with the place cut off for days by snow. As spring slowly failed to arrive, one of my first-year physics lectures dealt with the 'haar', the icy fog that often envelops the Athens of the North in that season. It appears when warm moist air sweeps over the cold North Sea and condenses into a fine mist that blows inland to soak its citizens. The phenomenon is well described in James Hogg's 1824 work *The Private Memoirs and Confessions of a Justified Sinner*, the inspiration for Stevenson's *Dr Jekyll and Mr Hyde*. In one scene the hero walks past Holyrood House, at the lower end of the Royal Mile, where 'the haze was so close around him that he could not see the houses on the opposite side of the street', and then climbs Arthur's Seat to find the summit in sunshine (to enliven the tale, his half-brother, who is possessed by the devil, emerges out of the dreary pit below and later murders him in one of the gloomy wynds of the Old Town).

Almost two centuries on, 'dreich' weather, if not demonic possession, still lays siege to the capital. The mist can last for days, and the atmosphere was in my time further improved by the scent of a nearby brewery, whose site is now occupied by the Scottish Parliament. In spite of the long days of summer (and I have happy memories of staggering home as the June dawn broke), Edinburgh has just twelve hundred hours of bright sunshine a year and in Europe only Glasgow, Reykjavik and Tórshavn in the Faroes do worse, while there are almost twice as many in the house at the edge of the French Pyrenees where much of this book was written.

I was lucky to get in to the University, for my A-level results were not impressive. A bad student blames his teachers, but many of mine were, in truth, dire. Wirral Grammar School was then, like many of its fellows, a watered-down version of what its masters imagined Eton to be, with gowns, houses, uniforms, canes, rugby football and all that stuff. Like its grand counterpart, it was devoid of members of the opposite sex, who were caged in a twin institution nearby and with whom we had, alas, almost no contact. The sole exception to its relentless conventionality was that the school song utilised the melody of the National Anthem of the USSR, perhaps in homage to our allies in the Second World War.

English, French and history were well taught, music and art appallingly so (the school's best-known alumnus is Kenneth Halliwell, who perhaps in reaction to his adolescent experiences there killed his boyfriend, the playwright Joe Orton, after each had spent time in prison for defacing library books to make collages). It had its post-war quota of disgruntled, disturbed, dismal and demonic teachers, balanced by a few memorable individuals. My Latin teacher, Ebenezer Titus Ebonorufon Fewry, was, we believed, the first African to teach in a British grammar school. For his pupils, he was a fearsome but inspirational figure. He once accused me, with some justification, of having a mind like a dustbin, and that description has comforted me ever since. Mr Fewry returned to Sierra Leone and rose high in the nation's legal establishment before his country fell into chaos. I worked for a time at the University there, and tracked him down to his home, a rusty shed at the edge of Freetown. He had become an old man and did not remember me at all, but we managed to exchange a few memories of Liverpool's Left Bank before I left. Battered as he was by his experiences, in a very African denouement he was later committed to a mental home, where he was murdered.

I recall other pedagogues for more banal reasons. In the sciences I had one excellent biology teacher, matched by a series of dreary and bored tutors in mathematics, chemistry and physics. Biology I remember for its trips to the Field Centre at Malham Tarn in Yorkshire, where quite by chance I collected a few of the snails that later became the subjects of my own research (and I later taught a couple of field courses there myself, with much of the time spent in failed attempts to hide my ignorance of basic natural history). However, when it came to the physical sciences my abiding recollection is of a two-hour session on the Earth's atmosphere – now, I realise, a fascinating topic – in which a dusty master droned a monologue that ended with the unforgettable statement that 'this explains why it only rains when there are clouds in the sky'. In time I abandoned the classroom altogether and spent my days in Liverpool's

Picton Library, an activity that gave me my only tie with the Beatles, for I sometimes walked past the door of the Cavern Club (about which I knew nothing) on the way there from the Pier Head.

My poor A-levels meant that my applications to the four universities of my native Wales were all turned down. I then discovered that Edinburgh had a later application date and dashed off a submission. An answer appeared almost by return of post: I could come. It was the most important letter I have ever received, for the city, and its climate, formed my career and to a large extent my life.

My first weeks passed in a haze of confusion and embarrassment. I met my first public-school boy (an experience from which I am yet fully to recover), and spent a night on Waverley Station when the commercial travellers' digs in distant Portobello to which I had been assigned threw me out because I played the guitar. We were addressed by the Principal of the University, the eminent physicist Sir Edward Appleton. He had won the 1947 Nobel Prize in physics for his part in the discovery of the ionosphere, Earth's shroud of charged particles, activated as it is by radiation from the sun. His work was one of the first steps in the study of the relationship of our home planet with its parent star, and of how, alone in the solar system, Earth has managed to keep both its oxygen and its water. He told the assembled youth that the happiest moment in his life had been when as a schoolboy he was introduced to calculus, and we were crass enough to snigger at what then seemed to us to be – but I now realise was not – an absurd claim. He is commemorated by the Appleton Tower, a structure of shameless ugliness built during my undergraduate years on the ruins of an elegant square as the first step in an extended campus that has destroyed large sections of the city's Georgian heritage.

In physical terms Edinburgh is, as Dr Johnson put it, 'a city too well known to admit description', and I will not attempt to do so. Grim as it might sometimes appear, the Scots capital was a wonderful setting in which to study, not just because of the excellence of its college's instructors but because it was a real place, with the University and its students woven into its fabric, rather than some academic Disneyland cut off from the world by walls of distance or privilege.

The University then still clung to remnants of its tradition as the 'Tounis College', whose job was to instruct local boys, who expected, and received, nothing more than education from their alma mater. In my day it owned little student accommodation and had no more than a minor role in health, in welfare, in careers advice and in the assessment of its customers' satisfaction, pastimes that now consume much of the energies of its staff.

In the half-century and more since my arrival, the relationship of the Town's College to its students has been transformed. In my first days, the challenges of being thrown in at the deep end forced a certain resourcefulness upon us all. That was matched by the institution's readiness to accept even its younger members as adults who could decide for themselves whether to sink or swim. As an undergraduate I was given a key to my own department, and I with many others often took advantage of that to work late in the library (and now and again to get a free bed for the night). Today, that idea would be unthinkable, and the place has multiplied its numbers by five times, houses many of its students, and has a network of support unknown in my time. In many ways that is welcome, but, somehow, part of its unique flavour has been lost.

Biology, too, was in those days in transition from what it had been – the study of animals and plants, from skeletons and stems to ecology and behaviour – to what it has become, an exercise in molecular anatomy, leavened with computer science. In my first weeks, we were told to go to the library and identify the most important paper published in the journal *Nature* in the previous decade. Somewhat bemused, I went to the stacks, and the answer at once became obvious. The spine of the 1953 volume was battered and grimy, two of its pages were covered with finger marks, and (the unforgivable sin) someone had underlined parts of it in ink. It was, needless to say, the Watson and Crick paper on the structure of DNA, the molecule that became the icon of the twentieth century. The vandal had highlighted the penultimate sentence: 'It has not escaped our notice that the specific pairing we have postulated immediately suggests a possible copying mechanism for the genetic material' (a statement which struck me even then as a triumph of false modesty). Much later I wrote a preface to a new edition of Watson's famously undiplomatic book *The Double Helix*, in which I faced the impossible task of bringing fifty years of genetics up to date in fewer than twenty pages. I made the perhaps tasteless joke that although Newton had seen further by standing on the shoulders of giants, Watson and Crick had preferred to stand on their toes. For that I got a stiff letter from Watson, but he redeemed that long afterwards with an invitation to his ninetieth birthday dinner in the Athenaeum (no false modesty there, of course).

I was, without fully realising it, taught as an undergraduate by some of the founding figures of genetics, a science that in my student days was little more than half a century old for it dated its birth to the 1900 rediscovery of Mendel's Laws, published, but ignored, thirty-five years earlier. Perhaps the most memorable was Charlotte Auerbach, a Jewish woman who fled Germany in 1933 and who came to Edinburgh to do a PhD. In the early years of the war, in what was then top-secret research, she discovered that mustard gas, already known to cause wounds that resembled those produced by high doses of X-rays, induced mutations in fruit-flies at a rate far higher than did radiation. That was the first step in what has become the enormous field of chemical mutagenesis, now central to the study of cancer and of DNA damage. As one of my more impressionable fellows put it at the time: 'It's like being taught physics by Newton!' and in some ways it was. Several others among my pedagogues were in at the birth of what are now sciences in their own right. They did fundamental work on the genetics of single-celled organisms such as malaria parasites, the control of cell division, the biology of fertilisation, the nature of embryonic development, the barriers between species, and more. All – unlike many of their modern equivalents – were happy to tell students about the latest developments in their field and, even better, to engage them in discussion over the vile but powerful coffee served in the basement canteen.

Many of my lectures, my undergraduate research project and my PhD thesis – still based in Edinburgh – had, although I did not realise it at the time, a tie to the work of Sir Edward and, in the end, led to a career, and to this book. Appleton, with many others, had studied how the upper atmosphere protects those on the ground from the rays of the sun. His work was a first step towards today's image of the sky above as a complex filter, without which life would not exist. Less than half of the radiation that strikes the outer atmosphere makes it to the surface on a sunny June day (and in the haar just a quarter as much), but those rays mould all our lives, physical, mental and – for some – even spiritual. My research, trivial as it may be in comparison to his, has done a little to show how they rule the lives of plants, animals, and ourselves.

The notion that our nearby star drives the weather, forms the landscape, fuels – but sometimes kills – the creatures that live upon it, controls almost all daily and annual patterns of activity, improves the mood of those who bask in its rays, and even acts as a seat of divine authority, can be traced back to the Greeks and beyond.

They, like many of their successors, were uncertain about the balance between its benefits and its dangers. Zeus, God of the Sky, destroyed his enemies with thunderbolts but also protected the harvest, while his son Apollo ruled the sun, poetry and music, but, when minded to, brought the plague. In Queen Victoria's day, too, the star was seen as a mixed blessing, for although it never set on her imperial possessions it was known to be dangerous to those who ruled over them. For women, exposure might cause a tan, a sign of social inferiority, while the opposite sex faced challenges of its own. An early twentieth-century volume entitled *The Effects of Tropical Light on White Men* warned that its rays led to irritability, lack of drive and of concentration, loss of memory and of appetite, headaches, insomnia, diarrhoea, depression, phobias, heart palpitations, excessive masturbation, insanity and suicide. The only protection was to wear red underwear.

Whatever its effects on intestinal health and autoeroticism, all the other ailments listed in that great work are indeed related to sunlight, but – as my own pages try to explain – to a shortage, rather than an excess, of its offerings.

A few years after its publication, attitudes began to change. Robert Baden-Powell, founder of the Boy Scout Movement, recommended in a 1922 work that 'young fellows in their rutting stage' should carry out outdoor exercise naked or with as few clothes as possible to allow sunlight to permeate body and mind and to reduce the ever-present danger of self-abuse. His suggestion developed into the healthy-living craze of the post-war years, with its emphasis on life in the open air – a dogma that persisted into my own adolescence, no doubt to my benefit. In recent years, however, the climate has altered again. Television, laptops and mobile phones mean that instead of the forcible ejection into the fresh air I faced in the boisterous 1950s, young people today spend, on average, an hour a day less outside than they did even a decade ago.

The general principles of the effects of sunlight on ecology, climate, geology, health and temperament were established long ago. However, since my own stint north of the Scottish border, each of those topics has been transformed. *Here Comes the Sun* tries to weave such advances into a single narrative and, as all I have to report of any interest about my own life is science, the occasional anecdote may creep in. To set the scene I preface this Preface with a brief account of some of the ways in which I have myself tried to explore a few of the talents of the solar furnace.

In a belated revenge on the Caledonian climate, much of my time since I last heard the cry of its east wind has been spent under blue skies. Like Stevenson, I have endured quite a long spell in California, but I never made it to Samoa (I did, however, get to Hawaii). A fifty-year search for illumination has also taken me across Europe, to the Americas, to Australia and to Africa. In my student days an interest in the tie between solar energy and biology seemed a minor specialism, but the subject has grown to encompass some unexpected parts of the world of science.

Whatever the local climate, the interaction of animals with their planet's source of energy influences their behaviour, their ecology and their internal machinery. In addition, it helps to maintain genetic variation and directs the course of evolution. My own research touches on all those themes. Perhaps the sole piece of general advice to emerge from it is to remind every scientist of the value of E. M. Forster's phrase 'only connect': always search for links of one topic with others from disparate subjects, in the hope that they will provide a hint about what to do next. My explorations into the perhaps arcane field of the thermal biology of snails and fruit-flies has had rather little impact on climatology, physiology, ecology and the rest, but at least it has been possible to make many links in the opposite direction.

In the past three decades technology has transformed the science of life. Genetics and evolution have become two faces of the same subject. The former – to use a somewhat hackneyed analogy – provides its vocabulary, while the theory of evolution gives the field its grammar.

In my student days, in contrast, the study of inheritance, of the history and geography of plants and animals, and of the flow of energy through bodies and through communities scarcely overlapped. Molecular biology had just begun, but the idea that one could read the language of the genes seemed unthinkable. Research on inherited diversity in nature was confined to the few creatures that gave a hint about what variation they possessed. They included fruit-flies, butterflies and a few plants. My own favourites – and that of many biologists of the period – was a certain group of land snails. Their attraction was (and is) that they have a system of inherited variation in shell characters that makes it possible to count genes in natural populations. I have collected tens of thousands of the creatures in an attempt to work out quite why such diversity is there, and have accumulated published and unpublished data on hundreds of thousands more (and Stevenson comes into the tale, for he once said: 'It is perhaps a more fortunate destiny to have a taste for collecting shells than to be born a millionaire'; I agree, although I do regret missing a legacy).

A couple of my Edinburgh contemporaries gained Nobel Prizes, but such distinctions are not offered for excellence in the study of molluscs. Even so, and even in the face of the ridicule faced by those involved in the subject, I do not regret the years spent in their company. Part of the reason is the simple pleasure of field trips to spectacular parts of the world, but more has to do with my role as a minor participant (and enthusiastic bystander) as that small corner of biology has been illuminated by discoveries in other fields.

Now, I am sorry to report, the advance of technology means that my favourite creature has lost its modest glamour. I used to be able to say that 'I am one of the top six experts on the genetics of snails and the other five agree' but now I doubt that even half a dozen of us are left. The cost of reading the message of DNA has collapsed. In the year of the Millennium, when the (then much-mocked) notion that it might be possible to sequence the human genome began to make serious moves forward, the price was estimated at a hundred million dollars per head. Now, that has dropped to around a thousand dollars, and may soon be even less. Why, then, study a system like mine, with a few hundred variant forms in shell colour and pattern, while in any creature, from bacteria to humans, ingenious machines make it possible to identify millions within the double helix itself?

Not just molluscs have moved out of the limelight. Once, most biological roads led to fruit-flies, for until not long ago almost every advance in genetics was made with their help. I too have spent plenty of time in their company. Now, even the glory of the fly has begun to fade, and *Homo sapiens* has replaced *Drosophila melanogaster* as the preferred raw material of much of the science of life.

In many ways that has been a positive move, for that species is a far more tractable subject for the study of genes in populations than is any other creature. Patterns of human diversity are better understood than are those of other animals, with hundreds of thousands of copies of the double helix read off across the world, while techniques to identify, and perhaps even to repair, damaged genes have begun to enter medical practice. The evolutionary tale told by DNA is supplemented by the physical and written records of the past, so that history, anthropology and the double helix have been welded into one. It has even become possible to extract the stuff from bones tens of thousands of years old (and in a tip of the hat to my favourites, snail shells can also retain their patterns long after death).

My own only direct contact with the genetics of our own species came with a trip to Israel long ago, where I took mouth swabs from citizens of Palestinian descent in an attempt to study their history of migration. We found that many families, as tradition told them, had lived in the same villages for centuries and were close kin with their neighbours.

On my way home, the security people at Tel Aviv airport became interested in the cardboard box in my hand luggage. What was in it? My reply, 'Arab spit', did not go down well, but when they discovered what it contained they perked up and asked whether a DNA test might distinguish those with Jewish ancestry from others. I did point out – as of course they knew – that once such tests had been used to murderous effect, but, in a reminder that human genetics raises ethical questions not faced by those who work on fruit-flies or molluscs, they seemed unconcerned (the answer to their question is, more or less, 'Yes').

The information that emerged from those samples became a very small part of the huge databank that now tracks our species' movements across the globe. That mass of figures also shows how humankind has been exposed to the Darwinian machinery in just the same way as have less pretentious creatures. Grey skies may lead to gloom, but they also explain why the natives of Edinburgh and Glasgow have evolved to become the palest people in the world (Billy Connolly, quoted at the head of this chapter, once claimed that his natural colour was light blue and that he needed a week in the sun before he turned white). Their pallor, discussed at greater length later in these pages, also lies in part behind the regrettable fact that the lives of Scottish men are two years shorter than are those of their fellows in Manchester or Birmingham.

And still, as biology raced onwards, I continued to plug away at my two favourite organisms. As is the case for many scientists, some of my research has been a modest success, while more has been a complete failure. If a theme emerges as I look back on the work, too much of it combined an excess of ambition with a shortage of common sense. One of the famous experiments in the evolutionary biology of my undergraduate days involved industrial melanism – the spread of dark-coloured genes in moths in smoke-blackened British cities. There was then an attempt to measure the intensity of natural selection by birds as they picked off the lighter versions from soot-covered urban trees and their opposites from the unspoilt trunks of the countryside. The work was (and to an extent still is) known to every student, but was flawed because the researcher pinned dead specimens where they would be most conspicuous, which much biased the results. Early in my career I resolved to do the job properly, with live snails and flies rather than dead moths.

That was somewhat of a blunder. In various places in Europe and the Americas we released tens of thousands of snails, and hundreds of thousands of fruit-flies. The main result was that we never saw most of them again, for we grossly misjudged the numbers already there, the ability of flies to migrate – or be blown – over large distances, and the talents of snails to stay out of sight. I did learn, to some extent, from my mistakes, but it was an expensive lesson. I skate over such failures here, and like most scientists, concentrate on the stuff that worked.

The most common species in my favoured group is referred to in natural history books (but as far as I can establish nowhere else) as the Wood Snail. Its Latin name is *Cepaea nemoralis* – and *nemoralis* does mean grove-dwelling – and the creature is abundant across much of Europe. The colour of its shell varies from pale yellow through dark pink to almost black, and its surface may be decorated with up to five dark stripes, which themselves come in several forms. Such differences are controlled by genes. With the help of friends (most of whom were not biologists but who came along for the ride) I have collected these animals and their relatives from the Hebrides and northern Norway to Croatia, northern Spain and Romania in the hope that I might uncover why the frequency of their inherited variants changes so much from place to place. I have also tried to ask the perhaps more ambitious question as to quite why the variation is there in the first place.

The job involved lots of travel. I did my undergraduate work on the genes of those creatures near their northern limit, in Scotland. One of the first trips was to the small island of Raasay, in those days a centre of the Free Presbyterian Church of Scotland (referred to by detractors as the Wee Wee Frees), the product of several splits in the Caledonian religious establishment and practitioners of the strictest form of Calvinism. They generously allowed us to stay in their disused manse, on the understanding that we were not to be seen outside on a Sunday, unless we wished to attend their service. None of us did; but it was cold and dull inside and we lit the fire. That in turn set the chimney alight, which meant that the worshippers had to break their own Sabbath rules to help put it out. We were not invited back.

My first expedition outside these islands, as I began my PhD work, was to the Velebit Mountains in Croatia. In the mid-1960s this was still a wild and woolly place with most of its roads almost impassable, but it had a spectacular limestone landscape of enclosed basins surrounded by rocky hills, and limpid lakes and rivers. I learned some basic Croat (one of my first phrases was *Nisam njemački*: I am not German), and over four annual visits of two or three months each I came to know the area well. No decent maps were available to the public, perhaps for reasons of security, but after the total disorientation suffered on our first trip I begged some military maps from American sources. I received them under strict instructions that I should not take them into the country itself (an order which, with a regrettable lack of responsibility, I ignored).

I also read with interest the *British Naval Intelligence 1944 Geographical Handbook of Jugoslavia* ('For the use of persons in HM Service only'), with its account of the nation's topography, climate and economy, together with a summary of the chaotic history of the Western Balkans. I was struck by the differences among the villages of the Velebit, some prosperous and some poor, some with tiled roofs, others with slates or corrugated iron. I also noticed the ramshackle state of many houses, with bare brick walls supported by solid stone bases (huge numbers had been destroyed at the time of the Second World War in a simultaneous battle against the invaders and a bitter civil war within Yugoslavia itself).

Even so, my insights into the country's hidden discontents in those last days of Marshal Tito were feeble indeed. Memories of the bloody internal conflict of twenty years earlier still bubbled under the surface, as was manifest in the advice we occasionally received from the inhabitants of one village about the dangers of going to the next – a concern based entirely on misplaced nationalism. The friction became all too clear years later when the local Serb minority in our study area clashed with the Croat majority as Yugoslavia broke up. Thousands were killed on both sides, and the small town of Gospić, close to where we had our own quarters, was hit by a barrage of shells, on one notorious day by as many as three thousand five hundred. In a 1991 massacre close to one of our old sample sites more than a hundred people, almost all them Serbs, were murdered.

I chose the Velebit not for its scenery or its history but because *Cepaea nemoralis* reached the southern limit of its distribution there. I found that the Balkan populations were far lighter in colour than their Scottish equivalents, and that observation has since been backed up by many samples from across Europe, with a close fit between the incidence of the genes that produce pale forms and the number of hours of bright sunshine.

So far, so simple; but those mountains were also the home of a related and then little-studied species, *Cepaea vindobonensis* (the 'Vienna Snail', but again always referred to by its Latin name), which has an attenuated version of the variation from light to dark found in its cousin. We soon uncovered large differences in frequency of the two forms of that species over a few hundred metres. In general, the light versions were found on the slopes, while the enclosed basins were the home of the darker variants. The reason why soon became obvious, for on many mornings the basins were filled with mist while the mountains were bathed in sunlight. The temperature in the frost-hollows was often as much as 5°C lower than that on the hillsides.

It did not take much physics to work out how such patterns evolved. As anyone who sits in shorts on a black iron park bench on a hot summer's day soon notices, dark objects heat up more in the sun than do light-coloured, because they soak up the solar output rather than reflect it. A simple check with a radiation thermometer showed that dark individuals of both species were several degrees warmer than their lighter cousins when placed in sunlight (although a two-year attempt to measure the strength of natural selection by climate on each form with the transfer of thousands of marked animals between mountains and basins came to naught).

As my research ground on, I extended the work on *vindobonensis* across the Balkans to see whether it showed the same large-scale changes in genetic structure as did its wood-loving relative. It did; but my main memory of the 1968 trip into Romania was the advice received from the British Embassy in Belgrade that we should abandon our plans because of the political situation. Just a few days before our intended arrival, Soviet troops had invaded Czechoslovakia, and there was real fear that Romania would be next, for a new (and still putatively liberal) ruler, Nicolae Ceauşescu, had just taken office. With juvenile overconfidence I ignored their counsel, but as I left the Belgrade offices I was handed a parcel and told to open it if I saw any signs of conflict across the border. Of course, we unwrapped it at once. It contained a large Union Flag which might – perhaps – have scared off the Russians. Fortunately we did not need it. Once in Bucharest, we were given a minder by the National Academy (who laboured under the misapprehension that we were an official expedition from the Royal Society of London; that organisation gave us the money to do the work but we had no such status, although I did not admit as much to our hosts). The minder had a heavily stamped piece of paper that caused consternation to those to whom it was shown, but opened all doors.

In the Balkans and across Europe (except on these islands where the idea that snails might be edible has never taken root), biologists who collect such creatures have an occupational hazard, the invariable question from passers-by: 'Do you eat them?' I know the answer – 'No' – in several languages but I never explain why.

I avoid them because I once suffered a surfeit of molluscs that turned mild aversion into real dislike. I had been asked to take part in a television programme that set out to explore the anomalies around the trade rules of Europe. The French had, it was claimed, planned to redefine the word *escargot* on a menu to mean not just 'snail', but 'French snail'; a creature born on Gallic soil.

The large Roman snail has long been the nation's favourite, but the species has been driven almost to extinction there. The smaller brown version, the *petit gris*, common in gardens, is tasty and easy to cultivate, but does not get its cousin's premium price. Some restaurateurs turned to a tactic that makes economic and ecological sense: to cook a *petit gris* body and insert it into an empty Roman shell, often one recycled dozens of times.

Then came the quarrel with Brussels. A resourceful Irishman who raised brown snails for sale in France decided to use a giant African species instead, for it grows to the right size far faster than the others. The French growers tried to put a stop to that with an import ban on the grounds that, like Parma ham in Italy, snails were a unique part of French culture. My small part in the tale was to visit a Parisian restaurant, L'Escargot, to report on any differences I might find between the French native and its impersonators.

Its owner provided me with examples of each kind, with generous amounts of wine. Unfortunately, the producer needed repeat after repeat of the same shot until, after a couple of dozen snails and an uncounted number of glasses of Macon, I made my excuses, left, and threw up. The experiment was a failure because I could not tell the difference among them anyway, but I have avoided all four ever since.

That is unfortunate, because plenty of animals that are, like snails, eaten by predators have a bitter taste that dissuades their attackers and bear warning patterns for protection. The bright colours of my favourite might, some have suggested, have emerged for the same reason. I once had a chance to test the taste hypothesis. In the Pyrenees we were offered specimens stewed with bacon, milk and orange peel and told that the yellow ones were the sweetest. I made an excuse and left, but some years later I tried to persuade a PhD student to do the experiment. He too quailed at the idea and it remains on my lengthy list of projects that might get done some day.

From Edinburgh I moved in 1970 to the University of Chicago, in January weather bleaker than the worst I had suffered in Scotland. The experience was somewhat of a shock. Any thoughts I may have entertained that I understood at least some population genetics evaporated at once. I had, almost by accident, been accepted into the most eminent evolutionary research group in the world. I had written, rather daringly, to its leader, Richard Lewontin, asking his advice as to where I might apply for a post-doctoral position, and within a couple of weeks received a letter saying that I should arrive as soon as convenient. By return I pointed out that I had not yet submitted my PhD thesis; he joked that this was just a case of 'post-doc, ergo proper doc', and I should come anyway (I did not in fact write it up for another five years: you could get away with such things in those simpler and perhaps happier days).

Chicago in the era of the Vietnam War was in ferment. The University is in an enclave called Hyde Park, which is surrounded by a poverty-stricken section of the inner city that suffers from a high crime rate. It protected itself with its own police force. I was rather surprised when, early in my time there, I went to a late evening study session in the library and saw that students were not allowed to leave except in groups accompanied by an armed guard, but I soon got used to it (post-doctoral research fellows, it appeared, were expendable and could be left to fend for themselves). I myself faced constant reminders about the need for caution, but never had any trouble.

When it came to research, my greatest weakness was, as it still is, my poor mathematics, and I remember the despair I felt when I went to the first of the weekly lab seminars of around twenty people and found not only that I could not understand the gist of the – highly mathematical – talk but that I did not even recognise some of the symbols on the blackboard (I tried to console myself that Darwin himself faced the same problem, but that did not help). I realised that my real job description was, like his, 'naturalist', rather than 'population geneticist', and blamed my school (unfairly, for I had missed plenty of later opportunities to catch up on the subject). Even 'naturalist' may be too kind, for Darwin, unlike myself and the vast majority of modern biologists, had a deep and intimate knowledge of the living world, while I could do little more than recognise a few molluscs.

Surrounded by theoreticians as I was I could, even so, make some contribution to the lab. A couple of years earlier, the Lewontin group had begun to use a simple technique called electrophoresis, the separation of biological molecules using a powerful electric current to drag them through a gel. How far any molecule travels depends on its size, shape and charge. They used it to reveal an unexpectedly high level of genetic diversity in fruit-fly proteins. The discovery opened the door to the now universal pastime of tracking human and animal movement, relatedness and evolution using genes. The laboratory used commercial apparatus that was slow, ineffective and badly designed. After a bit of tinkering I came up with a better, faster and cheaper version which, rather daringly, I then patented. A couple of years later I sold the rights to a British firm, and with the proceeds bought my first London flat (a venture now unthinkable for a young academic).

When it came to politics, Dick Lewontin's group was predictably radical in tone and rather disapproved of such financial manoeuvres. He himself was the first person to resign from the National Academy of Sciences, on the grounds that it had undertaken secret military research projects for the government and would not tell him what they were (rumour was that one had to do with designing a target strategy to bomb Haiphong harbour). I went on the approved number of anti-war marches, most of which were peaceful. One was less so, for the demonstrations after the lethal shootings by the National Guard at Kent State University of four student protesters against the bombing of Cambodia ended in riots. Discretion outweighed valour, and I observed the events at a distance, as I knew I would lose my visa if I got into trouble.

The main influence of Chicago on my scientific as distinct from my personal career was to shift my interests to creatures that fly rather than creep, and towards biochemical variation rather than that visible to the naked eye. Within a few months of my arrival, I found myself, rather to my surprise, in bright sunshine and temperatures that could be lethal. I was in a roadless desert on the Baja California peninsula of northern Mexico. The trip (in a light plane that landed and took off from the beaches; for research grants, those were the days) was in search of specialised fruit-flies adapted to life in different species of cactus, each plant with its own chemical insect repellent, but each botanical defence evaded by its fly's ability to evolve faster than its host. As I laboured beneath a searing sky I was astonished by the larvae's ability to survive inside rotting branches whose temperature could reach over 40°C, far higher than that tolerated by most snails or by the familiar *Drosophila melanogaster*, and a level lethal to many animals.

That observation led to an interest in thermal stress in fruit-flies that has lasted to this day. I spent several years commuting between London and Death Valley, in an extended search for genetic variation in the structure of proteins in flies from climates as different as the snowline in the Sierra Nevada and the salt flats of the Valley itself, the hottest place in the world.

On one of those trips I had another encounter with the roots of modern genetics, for I met one of its founders, the Russian evolutionary geneticist Theodosius Dobzhansky, then in his seventies. He had worked in Central Asia on the genes of wild horses and of ladybirds and butterflies, but came to the USA in the late 1920s and carried out pioneering research on fruit-flies across North and South America. He was proud of his distant kinship with Dostoevsky, but was even more so of having been given, as a young man, a cigar by William Bateson – the Cambridge academic who invented the word genetics, just after the rediscovery of Mendel's work – on Bateson's visit to St Petersburg.

Vladimir Nabokov's novel *Pnin* is based on a Russian exile who speaks heavily-accented English, who quarrels with all his colleagues and is forced to move from his beloved New England university when he has a falling out with a libidinous lepidopterist. The character with his eccentricities was Dobzhansky to a tee (and Nabokov himself was a keen collector interested in the genetics of butterfly hybrids, and the two corresponded and may well have met). I came across him on a field trip to the Anza-Borrego Desert in California. He had been ill and depressed, and on the drive there I sat somewhat overawed in the back seat as he talked to a fellow professor not about genetics but of his fights with administrators and of how his own work had never been appreciated. Once at the site, however, the warhorse heard the trumpets (or at least smelled the rotten tomatoes used as fruit-fly bait), and shed thirty years, hopping joyfully across the rocks. It was his last field trip, for he died a few months later; but it was both a memorable brush with greatness and a now welcome reminder of the ability of science to rejuvenate its more aged practitioners.

In the sample we collected in Anza-Borrego, and in many others across the deserts and mountains of California, we found plenty of variation in the structure of proteins, and some genetic changes from place to place, but experiments with marked flies showed that migration over many kilometres seemed to even out any effects of climate. That in turn led to another failed scheme in which I introduced flies from the cool mountains into the desert sites to test whether their genes could cope with extreme heat, and in effect never saw any of them again. We did get shot at on occasion by the eccentrics and survivalists who lived in some of the remote oases, but either they were poor marksmen or they just wanted to scare us off. It always worked.

Soon after my return to London from Chicago I found a job at University College London, and I have been there ever since (as I say to my young colleagues: 'Don't worry; the first forty-five years are the worst'). What, many people have asked me over the years, is the point of research on the thermal lives of flies and snails? Why should taxpayers subsidise such frippery? A few years ago, in fact, they stopped. For a while, through obscure routes, I was able to subsidise my fieldwork with cash that came, for example, from appearances in television advertisements for cars and for life insurance, the latter enlivened by a joint performance with Buzz Aldrin and Muhammad Ali, but soon those sources dried up too.

I continued to fund my work for the reason that drives every scientist: it was fun, albeit fun often disguised with grander terms. I am the last to claim that our results have much direct application to human happiness and prosperity, but research on such topics has at least directed science's attention towards our own relationship with the heavens, often with unexpected results.

Interest in that subject goes back a long way. The Swedish botanist Carl Linnaeus, the first classifier of life, was so impressed by our dependence on the sun's rays that he recognised humans as members of a group he named *Homo diurnus*, which he divided into three species called *Homo sapiens*, *Homo monstrosus* and *Homo ferus*, the wise, monstrous and wild varieties of the daytime primate.

In retrospect, he was right. Our lives, we now know, often bow to the sun gods. Those deities, at least in their physical form, affect bodies and minds, health and disease, and are even responsible in part for the ability to stand upright. *Here Comes the Sun* tries to show how they are important not just for *Homo diurnus* but for biology as a whole, and, in a wider context, for much of the world we see around us.

Robert Louis Stevenson's advice to writers was that 'there is but one art – to omit!', and the empire of the sun extends so far that I have omitted a great deal. This volume is necessarily broad, but inevitably shallow. Its early pages explore the four classical elements of earth, air, fire and water in relation to their parent star, and it moves on to other sun-related themes such as sleep, dreams, memory and mood.

The first chapter gives an account of our local nuclear reactor and its effects on our own planet. Its output rules the physics and chemistry of our home as much as it does the lives – and sometimes the deaths through heatstroke – of its inhabitants, from snails to humans.

For snails, fruit-flies and people, its rays bring lethal challenges, but also the gift of life. They work much of their magic through the medium of water, the agent of many of the sun's activities. Heat and rain together make the weather, which forms mountains, planes, deserts, lakes and rivers. Plants use sunlight to trap carbon dioxide from the air and to make the molecules that make us all. The flow of its constituents from plants to animals, men and women included, drives the world's ecology and, in the end, large parts of its economy.

Sunshine also has a direct effect on human welfare, for a shortage leads to diseases from rickets to multiple sclerosis and cancer. Dawn and dusk drive life's rhythms. Sleep, once the most mysterious of all the body's talents, is the key to many of its functions, memory included, and for those compelled to keep awake for long enough, madness followed by death is certain. In recent years, many people have begun to obey their own clocks rather than those of the solar system. As a consequence they face a variety of unpleasant side-effects, from obesity to short-sightedness and worse.

A shortage of sunshine, as I myself on occasion realised in my years in Scotland, can also lead to despond. James Boswell was born in Edinburgh, and spent many years there as a lawyer. He worried about his fits of depression – what he called his 'dispiriting reflections on my melancholy temper and imbecility of mind' – a weakness that manifested itself in his obsession, even as a schoolboy, with seeing public hangings in the Grassmarket a few hundred yards from his home. The weather made matters worse, and the winter entries in his *Edinburgh Journals* often refer to his mental torments. In January 1776, for example, he wrote that 'the thoughts of my own death or that of my wife or of my children or my father or my brothers or friends made me very gloomy'. A year later he complained of 'a morbid state of mind', and as Januarys passed he talked of his mood as 'low-spirited and languid', 'very wretched' and worse.

This problem has not gone away. Even in my time there, at examination season (and Caledonian education then revelled in tests in what seemed like every month) assiduous students in winter lost track of the hours and lived in almost perpetual darkness, with, for some, woeful (albeit temporary) effects on their emotional state. The long season of twilight still challenges many of its inhabitants. In Scotland the use of antidepressants has gone up by half in the past decade, while the nation's suicide rate, albeit in slow decline, is still a third higher than in its neighbour to the south.

The great biographer's attacks of despair lessened after his move to London, and Stevenson, too, noted how his own mental attitude improved after his escape to the Pacific.

Many other people have found that a few hours spent under blue skies generate a sense of well-being after a cheerless winter – a sentiment I still savour as I set off southwards each year (and science has begun to reveal the chemical changes that take place in our bodies when sun strikes skin to make a cocktail of the hormones that cheer us up). Far more people can now afford to indulge in that biochemical experiment than in my student days, for in the year I left school the British made around three million trips overseas, most of them to the shores of the Mediterranean, while today that figure has gone up by twenty times.

This book gains its title from a famous track on the Beatles' *Abbey Road* album, and is published close to the fiftieth anniversary of its release in 1969. George Harrison, who composed and sang 'Here Comes the Sun', noted that:

> It seems as if winter in England goes on forever, by the time spring comes you really deserve it. So one day I decided I was going to sag off Apple and I went over to Eric Clapton's house. The relief of not having to go see all those dopey accountants was wonderful, and I walked around the garden with one of Eric's acoustic guitars and wrote Here Comes the Sun.

The spring of that year was indeed one of the sunniest ever seen, while the previous winter had been harsh, with heavy snow even in December.

Today the author of that celebrated air, like my undergraduate self, like Robert Louis Stevenson a hundred years earlier, and like James Boswell a century before that, would have less reason to complain about the weather. Over the past five decades *Homo sapiens* has burned vast quantities of ancient sunlight in the form of coal, oil and gas in an attempt to return to a diluted version of the balmy tropical landscape in which he evolved. Almost everyone in the developed world, the people of Edinburgh included, now lives on an allegorical savannah, a carefully constructed replica of our ancestral home, a place where the season is always springtime. Technology has taken us at least part of the way back to Africa.

To pay for the journey we have squandered an ancient investment in biological carbon. A typical city dweller now needs twenty times more fuel to keep at a comfortable temperature than would a wild mammal of the same size, and the average Briton spends a substantial portion of his or her income on ensuring that the thermometer stays at an agreeable level. Across the world the desire to heat or cool houses and offices in an attempt to re-create an ideal internal climate accounts for almost as much fuel use as that consumed by industry or by transport.

Welcome as the new and artificial climate may be, Nature is not mocked. As we burn fossil fuels, their carbon is released into the air, and the planet has warmed and continues to do so. As it does, mankind has been reminded of the age-old authority of sunlight over the Earth's affairs as temperatures rise, deserts advance, rivers run dry, and sea levels continue their inexorable rise. In its final chapter, this book discusses the biological and physical changes that may come from our disregard of the fragile relationship between our star, its third planet, and those who live upon it.

Not all the news is bad. San Francisco has seen a reduction in coastal fog of about a third in the past century, and the same is true in other places. If eastern Scotland follows the trend, the horrors of the haar will fade away, and perhaps one day Edinburgh will – as Goebbels hoped – make a delightful summer capital for Britain's citizens. When it does I will move back at once.

CHAPTER 1

THE EMPIRE OF THE SUN

Everything has a natural explanation. The moon is not a god, but a great rock, and the sun a hot rock.

Anaxagoras, fifth century BC

Every second, our local star puts out more energy than man has generated since he learned how to light a fire. The descendants of those million-year-old flames now fuel cars, aircraft, factories and central-heating boilers. As they do they consume an annual four hundred million tons of black gold, the product of ancient sunlight.

A square set out in the Sahara Desert with sides three hundred kilometres long (the distance from London to Liverpool) would represent just one part in a thousand of its area. Covered by solar panels, that modest acreage could run the modern world for ever, and would put the coal, gas, oil and nuclear industries out of business. It might be expensive to build (though less so than the bailout of the banks after the 2008 financial crisis), but would involve almost no further expenditure. If one ignores for the moment the minor issue of how to move its output to where it might be needed, the problems of man-made climate change, air pollution and radioactive waste would be solved at a stroke.

That utopian scheme is not about to come to fruition, but it hints at the power of sunlight. Impressive as the figures are, they do not do justice to its wider talents. The solar system is an empty place. The Astronomer Royal Sir James Jeans, in his 1931 classic *The Stars in their Courses*, had a vivid image of just how deserted it is. He imagined it to be reduced to the size of Piccadilly Circus: 'If we want to make a model to scale, we must take a very tiny object, such as a pea, to represent the sun. On the same scale the nine planets will be small seeds, grains of sand and specks of dust . . . The whole of Piccadilly Circus was needed to represent the space of the solar system, but a child can carry the whole substance of the model in its hand.' The sun itself represents 99.86 per cent of the mass of the solar system, while the Earth has one part in a third of a million.

Our home planet, as seen by a (well-insulated) observer on the solar surface, would be no more than a speck on the horizon. It receives just two-billionths of the star's output at that distance. At the edge of Earth's atmosphere, about fourteen hundred watts per square metre flows in, enough to run a powerful microwave. That figure is reduced to around a thousand such units after the rays have fought their way through the atmosphere – itself, when scaled down, no thicker than a layer of varnish on a school globe. The input of the sun to its outer edge is equivalent to that produced by a hundred and fifty million large power stations, and without that protective atmospheric layer the surface would be bombarded by lethal radiation.

Newton split sunlight with the help of a prism, to give the first hint that it contained waves of different lengths. As he said, the rays have no colour; our brain processes them with the help of receptors for three different wavelengths that we interpret as red, green and blue.

The sun's waves stretch rather wider than that. Most of the sun's output has a wavelength of between 200 and 2000 billionths of a metre. That includes the visible range, which stretches from 400 to 720 of those units, infrared (longer than 720) and ultraviolet, which stretches from 2 to 400 billionths of a metre. Ultraviolet, UV, is divided into three bands, UVA, UVB and UVC, the first of which penetrates the skin and can cause sunburn, while the second generates a protective tan and the last does not make it through the atmosphere. UVB makes up about a twentieth of the total, and is a tiny fraction of the solar input as a whole, but is, except for visible light, biologically the most important part of the whole spectrum, for it generates a vitamin that is in today's crepuscular world often in short supply. Visible light, infrared and ultraviolet together represent about half the total that reaches the ground.

The capital of Ecuador (a nation named for its geography) sees more of that than does its Scottish equivalent, for in the tropics the sun at noon is overhead and each sunbeam heats a small piece of territory. In the far north or south, its rays have a reduced angle of contact with the surface, which increases the area illuminated and reduces their intensity. The output is further dissipated because it has to pass through a thicker layer of air in Midlothian than at the equator. Nearer the poles, the seasons also have an effect. In London, at noon on midsummer's day, a metre stick will cast a shadow about half a metre long. In midwinter it will (on the rare occasions when the sun is shining) stretch six times further. The deadening effects of the atmosphere on the vital rays mean that Bordeaux, halfway between the North Pole and the equator, gets little more than three-quarters of the UV dose that reaches Quito, Edinburgh receives about half as much, while Lapland gets even less.

The solar factory has several other products on offer. Neutrinos travel at the speed of light. Sixty-five billion strike our planet every second, but almost all pass through without stopping. Even so, those that do not make it to the other side make a tiny contribution to its donations. X-rays and gamma rays (some which have a wavelength of less than a billionth of a metre) pound the outer atmosphere, but a good proportion is blocked before they reach the surface. Those fortunate enough to spend time in the International Space Station gain a further insight into the protective effects of the air below them, for they may notice bright flashes as solar particles pass through their eyeballs.

As the cosmonauts admire the Earth beneath they see their home planet framed against the blackness of space. Earthlings, in contrast, as they look upwards in daylight, cannot pick out the observers as they circle above, but see blue sky instead. The colour can be blamed on dust. Blue light, with its short wavelength, is scattered to a greater extent by its tiny particles than is red, so that we perceive the noonday sky to have that hue. Close to sunrise or sunset, the sun is so low on the horizon that its rays must pass through a layer of air twenty times thicker than that traversed at midday. Even more of the blue light is then scattered out into space, so that at dawn or dusk an observer on the ground perceives more of the red.

Whatever its colour, the bright light in the heavens was for much of history seen as a deity rather than as a physical object. Those days are over. Modern solar science began with Copernicus, and by the middle of the twentieth century had revealed a great deal about its surface and its chemistry. Today's satellites and electronic probes say much more about what goes on inside.

The star is just over four and a half billion years old, with a radius of around seven hundred thousand kilometres and an internal density eight times that of gold. Impressive as it might seem to our eyes, the sun is no more than a modest member of the ten million million such objects in our galaxy, and sits at the centre of a smallish solar system. Its distance from Earth gives us the unit of cosmic measure known as the 'astronomical unit', which has been defined with uncanny precision as 149,597,870,700 metres. (Exact as that sounds, the unit is, given the annual and long-term variations in our planet's orbit, a somewhat abstract concept, as it is based on the average of today's maximum, and minimum, distances from its star.)

A century and a half after Newton, the German astronomer Joseph von Fraunhofer noticed that the solar spectrum was interrupted by a series of dark lines at irregular intervals across the otherwise continuous output. Each one, he realised, marks the absorption of light of a particular wavelength. They coincide with that of the light emitted by every element when raised to a high temperature; sodium, for example, shines yellow, and copper green. The lines send a message from the sun that makes it possible to work out its chemical composition without the trouble and expense of having to go there.

The solar furnace contains about the same range of elements as does our native planet, but in quite different proportions. Hydrogen makes up just under three-quarters of its mass, helium a little less than a quarter, while the others (sodium and copper included) make up the rest. That arrangement dates back to the turbulent era when the solar system was born from a disc of dust, the remnants of the explosions of dying stars into supernovas. For a short time every member shared much the same composition, but because the sun was so massive it soon attracted the lightest components – hydrogen and helium – of its neighbours, while they, with their limited gravity, could hold on only to their heavier elements. The power of that force is still manifest in our own planet, with its own heavy core of molten iron surrounded by less massive rocks, with almost all its water in the seas and most of its oxygen and other gases in the air.

The sun is not, as most people know, a hot rock. The ancients saw the radiant mass as a fiery chariot, but it has no external fuel and no oxygen. For a time, it was thought to be made of molten iron. The lustrous disc does look like liquid metal, but its temperature is in truth far higher than that in any foundry.

A more recent idea was that it shone because of a constant rain of meteors on to its surface. That belief too fell by the wayside. Then came the claim that it was a nuclear reactor, in which heavy elements broke into lighter ones and generated heat, as had the Hiroshima bomb. That notion was plausible, but wrong.

The sun is indeed a reactor, but one far mightier than any made by man. It does not split atoms, but unites them. When, under extreme heat and high pressure, nuclei are forced close enough together, they fuse. As that reaction takes place they lose subatomic particles such as protons and neutrons. The mass of the new element that emerges is then a little less than that of its parent, and a large amount of energy is released as a by-product. The sun does its job when hydrogen atoms join to form helium.

Artificial fusion was achieved for an instant on Earth with the first hydrogen bombs (although lightning flashes have long done the job for free). The idea that it could be used as a source of electricity that might be infinite and involve minimal cost has been a will-o'-the-wisp ever since. In 1954 the Chairman of the United States Atomic Energy Commission claimed that one day American children would enjoy electricity 'too cheap to meter'. His forecast was a hint at the practical potential of fusion, then still a state secret. His generation's children, their grandchildren and their great-grandchildren are still waiting.

In principle hydrogen – in practice, deuterium and tritium, heavy forms of the stuff – could indeed be confined within a magnetic field and squeezed until their atoms are forced together until they pump out more power than is needed to hold them in place. The Massachusetts Institute of Technology built a machine that worked for two seconds and generated no external power. It continues to promise, as it has long done, a breakthrough within fifteen years. A vast international experiment in France has the same aim and is not much closer to fruition, with the first attempts at fusion planned for 2027. The hydrogen bomb in the sky has been busy at the task since it was born, and as an incidental has brought light to our lives. It might make more sense to use solar cells in the desert to trap the waves that pour in from space than to try to build a puny and expensive version of the same technology closer to home.

Nuclear fusion is at the heart of all stars until they reach extreme old age. On their deathbeds some among them explode into flares energetic enough to fuse even heavy elements together, which is how iron and the like emerged in the first days of the universe. Our own source of radiation is feeble, for it can do the job just for the most lightweight elements, hydrogen and helium. At its demise our sun will expand to become a red giant, which will engulf our own planet, before it shrinks to emerge as a white dwarf no bigger, but far denser, than the Earth, consumed as it had been by the dying colossus long before.

Light, as everyone knows, is made up of waves, but in the eerie world of quantum mechanics that is not the whole story, for it can also be represented as a series of minute particles called photons. A ray of sunlight is built of vast numbers of these particles, and the wave represents an average of their distribution as measured on a larger scale. The particles fire up our visual system, enable plants to make oxygen, activate solar panels and fuel the weather. An individual photon has a tiny impact, but their total input is impressive.

It takes photons eight minutes and twenty seconds to get to the Earth once they leave their birthplace – but before they can make that brief trip they must spend an average of some two hundred thousand years in its fiery core.

The fusion of hydrogen into helium becomes possible at a temperature of around 15.5 million degrees. Only the central fourth of our star, a sphere with a diameter thirty times that of Earth, passes that threshold. Each fusion produces a tiny quantum of energy, but together they fuel the solar reactor.

Every second, six hundred billion kilograms of hydrogen are forced into its machinery, and four billion kilograms less of helium emerge at the other end. That mislaid matter shows itself as sunshine. Every second, the fusion reactor shrinks a little, but that is no cause for concern, for it has enough hydrogen in reserve to let it light up the skies for another four billion years before it runs out of steam, or protons.

When viewed through a telescope, the solar surface resembles that of a pot of porridge, with surface plumes or 'granules' that may be thousands of kilometres across. They are indicators of internal upheavals. Towards the equator dark sunspots that change their position from day to day often appear. They were first noticed by Western scientists in the seventeenth century, but – as so often – the Chinese had noted their presence long before, when they observed the solar disc through a veil of cloud. Such structures come in bands, north and south of the solar equator, and vary in number in a cycle that lasts for around eleven years (and at present is on a downward curve to almost zero from its last peak of almost two hundred in 2014). For seventy years in the fifteenth and sixteenth centuries none at all were recorded, and similar gaps took place in the distant past.

Some claim that such events lie behind periods of cold weather on Earth, such as the 'Little Ice Age' in the seventeenth and early eighteenth centuries, but the evidence is not persuasive. Other supposed correlations include the density of porcupines in Quebec over the past century, and an 'index of mass excitability' that goes back to the Battle of Hastings and blames wars on peaks in the number of spots. In the nineteenth century a statistician – based, I am sorry to say, at University College London – also claimed to find a fit between commercial activity and the solar cycle. A closer look showed that to be mere coincidence and gave rise to the term 'sunspot economics', the mistaken (but widespread) belief that some in fact unconnected variable is behind movements in the market, or for that matter the climate.

Sunspots appear to move across the solar disc. That observation was the first hint that our local star, like all the planets that surround it, rotates. It does so in a manner quite different from that of our own familiar shift between day and night, for its outer layer is liquid rather than solid, and different parts spin at different rates. Solar blemishes near the equator move faster than those further away, proof that the waist of the star moves at greater speed than do its poles. The surface also vibrates like a deep-toned bell and responds to low-frequency waves – sounds, with nobody to hear them – which are emitted as a constant rumble as the active centre rotates and grinds against the surface layers that themselves rise, sink and turn.

The internal economy of a pot of porridge is simple enough, for its contents bubble all the way through. In the sun, in contrast, just the topmost fifth indulges in convection. The innermost mass consists not of porridge but of plasma, a mixture of electrons and ions that appears when electrons are freed from their host atoms, to generate a form of matter that stands apart from the more familiar solid, liquid and gas.

As the designers of the hydrogen bomb were well aware, the gamma-ray particles that emerge from a fusion event are potent and aggressive, and can cause immense damage. The radiation that reaches Earth's atmosphere is, in contrast, pretty tame. The dangerous products of the solar fusion reaction face a severe re-education within their birthplace before, battered into submission and with a marked change of personality, they are allowed to leave.

Light and its relatives travel at a uniform and high speed when in a vacuum, but slow down when they pass through glass or other dense substances. The sun's centre is so dense that the photons that emerge from a fusion event take two hours to cover even a metre. They must then traverse a straight-line distance of half a million kilometres before they reach the star's surface, a task made more burdensome because they have no sense of direction and just zigzag back and forth until, at last, they escape.

Inside the sun, the new-born photons enter a vast electronic maze. They hit an obstacle almost at once, and in effect bounce off it, hit another one, and then another. Each photon is like a drunkard who as he tries to find his way home bangs into an almost infinite series of fellow revellers along the way. Each endures trillions of collisions until, many years later, it reaches the exit and staggers out. Rather like those who have spent too much time in the pub, they start out in an aggressive frame of mind, but each time they hit an impediment they lose a little of their bottled-up vigour. Bit by bit, they decline into a more amiable middle age. First the gamma-ray particles become X-ray photons, and as they continue to lose their clout the majority turn into even milder ultraviolet elements. As they leave the plasma each, on average, retains no more than one part in several thousand of its infant bad temper.

Then they reel out of the pressure cooker and into the porridge. As soon as they find an up escalator in the convection zone, life at once gets faster, cooler and even more obstructive. The photons must now contend not just with their fellows and with hydrogen and helium as impediments on their journey, but with heavier elements – metaphorical lamp posts – that can maintain their identity in the new and cooler conditions. As the photons bounce off, the obstacles further soak up their animal spirits, so that when – at last – the minute particles reach the edge of the sun, ready to launch themselves into space, the majority have been reduced to shadows of their earlier selves, within the visible and the infrared range, with no more than a few survivors of the fierce creatures that all had been in their youth.

Towards the end of their journey, the photons enter the bright light zone we see from Earth. This is no more than five hundred kilometres thick, and is far less dense than what lies below. There they undergo a final trial by ordeal, and at last set out to find a life in the outside universe. A tiny proportion among them make it to the edge of our atmosphere.

Sunspots are evidence of another product of the solar furnace. They travel in pairs, because they are points at which twisted lines of magnetic force escape from the interior and flash in huge arcs across the surface. One spot represents the north (or positive) and the other the south (or negative) pole of a magnetic storm. The lines that link the two as they cross into and out of the surface are twisted and tangled. They block the passage of the hot central material as it bubbles up to produce a dark and relatively cool spot. The chains move across the surface and within the deeps of the sun as the star convulses under its storms of magnetism.

The sun is a gigantic dynamo. Some of its energies are dissipated as the 'solar wind', an ionised plasma that streams out at high speed and takes two to four days to reach us. Much of its output is blocked when it meets Earth's own magnetic shield (of which the layer identified by Sir Edward Appleton is a part). It does now and again leak in at the poles and spark off displays by the Aurora Borealis and its southern equivalent, but these do no harm. Their shimmering red, green and violet colours are generated when bursts of electrons collide with nitrogen and oxygen atoms.

Now and again, though, the chains of force on the solar surface are wound too tightly, and a distorted mass of material pops out to relieve the tension. That generates a local magnetic hurricane that may be visible from Earth as a flare on the solar surface. The radiation from such events can cause real problems. The Moon lacks the aerial buffer that protects its parent, which would be bad news for future visitors caught out in an electronic tempest. Even frequent fliers on Earth are at risk, most of all those who cross the North Pole, because five times more cosmic rays get through the protective layer in that region than they do near the equator. On each such trip, every passenger is exposed to the equivalent of two medical X-rays, or – should they be unlucky enough to encounter a magnetic storm – even more. Five return journeys a year may take them to the radiation limits set as safe by the International Commission on Radiation Protection. Aircrew make such journeys dozens of times over that period, and in the United States are classified as radiation workers, for they receive several times the annual dose permitted for X-ray technicians. Transatlantic pilots who have been in the profession for many years suffer double the average rate of cancer. They shrug off all calls to wear radiation badges.

Some solar storms are fierce enough to fight their way through our own atmosphere even in places far from the poles. Their charged particles create havoc as they generate surges of current on scales from microchips to national grids.

Radio messages, navigation instruments and electrical networks all suffer. The largest ever recorded was in 1859, the year of *On the Origin of Species*, when a huge flare was seen to burst out of the solar surface. This was followed by a spike in the magnetic recorder in the observatory that then stood at Kew, and by a spectacular aurora seen as far south as the Caribbean. Large parts of the telegraph system, by then in widespread use, failed because of electrical surges, some large enough to set fire to the equipment. In 1989 it was Canada's turn to pay the price. Its northern location and long power lines put the country at particular risk. In March of that year another great storm struck. Within minutes, Quebec's electricity network was shut down. It took weeks of work and millions of dollars to put right.

A repeat of such an event in today's connected world might drive a modern economy into meltdown. America's Homeland Security Committee estimates that an event the size of that in Quebec 'could pose the risk of the largest natural disaster that could affect the United States', with a chance of seven in ten that such a cataclysm will happen within a century. Many electrical systems and internet links are now protected from solar storms, but satellites are harder to shield, and a severe storm would overcome even well-protected systems on the ground. A return of a Darwinian hurricane might be catastrophic, with damage in the United States alone forecast to cost more than two trillion dollars. On 23 July 2012 there was a solar eruption as great as that event, but fortunately this was a near-miss. Had it happened a week earlier, our planet would have been in the firing line, there would have been a direct hit, and the world economy would, in effect, have been hurled back into the eighteenth century.

Human recklessness has already hinted at what might happen when a powerful electrical tempest arrives. Atmospheric tests of nuclear weapons came to an end in the early 1960s. A few months before they were banned, the United States military detonated bombs at the edge of space, just to see what might happen. The biggest of all was the 1962 Starfish Prime Test. A bomb a hundred times larger than that used at Hiroshima was launched just before midnight, from the remote Johnston Island, part of the Hawaiian chain.

The explosion caused an electromagnetic pulse, but one far greater than expected. It damaged electrical circuits in Honolulu and blacked out hundreds of street lamps and telephone lines. The sky lit up with a bright aurora (a glow had been forecast, which had tempted some hotels to offer 'rainbow bomb' roof parties in celebration), but again the display was more spectacular than planned, for it covered thousands of square kilometres and lasted for several days. The radiation belt circled the Earth and knocked out seven satellites (one of which was Britain's first venture into that field, the satellite Ariel 1, part of the research programme initiated by Sir Edward Appleton). The pulse travelled furthest when it followed the natural magnetic lines of force, with a parallel burst of bright light at the equivalent point on the other side of the globe. The Soviet newspaper *Izvestia* headlined the event as 'Crime of American Atom-mongers: United States Carries out Nuclear Explosion in Space'. Three months later, the Soviets staged their own experiment. They were rewarded by the destruction of hundreds of kilometres of telephone lines, and a fire that gutted a power station.

Below the magnetic shield, the Earth's atmosphere is further divided into sections, defined by their relationship with the solar output. In the lower atmosphere, or troposphere, temperature drops with height. That segment is about eighteen kilometres thick at the equator, but less than half that at the poles. Above it lies the stratosphere, a rarefied layer whose temperature increases with height as its gases soak up radiation. About eighty kilometres up, the ionosphere – generated when cosmic rays strike atoms of carbon, oxygen, nitrogen and more – keeps out much of the harmful radiation and, as Marconi and his successors discovered, allows radio waves to bounce back and forth to the ground, and like a stone as it skips across a lake, to travel great distances.

Not just displays of solar (or military) bad temper are blocked by the atmosphere. Radiation in the visible range is also kept out. One-third is reflected back into space, most of it by clouds and by dust. A further sixth is absorbed. A grey sky cuts out around half of it and a thunderstorm even more. On a day of intermittent sun, the clouds themselves look white, for they return light reflected from the surface, which means that some patches of ground get a bigger dose of photons under such conditions than they would under an open sky.

Once the sun's waves have fought their way through the air, clear or cloudy, they face another challenge when they hit the surface, for not all of them are absorbed. The proportion bounced back, whether from the surface or from the tops of the clouds, is measured with a statistic called the albedo, whose value varies from zero to one (I was once told, perhaps as a joke, that this was named after a medieval Arabic astronomer; I believed that for a while, although the word is in fact the Latin for 'whiteness').

After his sojourn in California Robert Louis Stevenson spent months on a cruise around the Pacific looking for the ideal place to settle down. In time he chose the tropical paradise of Samoa, with its white coral beaches. From there he made several visits to Hawaii. Most of that island's beaches are black, because they consist of ground-up lava. A coral strand absorbs no more than a quarter of the sunlight that falls upon it, but its volcanic equivalent soaks up nine-tenths. Conditions may then grow intolerable for the hopeful Hawaiian sunbather, or even for someone who wants to make it across the black sand to the sea. That contrast is a reminder of the importance of colour in the thermal relations of the Earth and those who live on it.

Fresh snow has an albedo of almost one (which is bad news for Antarctic explorers, for they face simultaneous sunburn and frostbite). That of the ocean is close to zero. As a result it soaks up more than half the solar radiation that hits the surface and becomes the main storehouse of its heat, most of it held in the top metre of water. Without Neptune's help, climate change would have reached crisis level long ago.

The several sciences of the sun have been transformed in the past five decades. More and more of what makes our lives, those of other creatures, and the world we live in, can be traced to its effects.

That idea resounds through history and, from Stonehenge to Mayan pyramids, almost every culture has had monuments to its sun gods. The solar disc does indeed have many of the attributes expected of a divine being. It has brought life to our planet and light to our lives, it regulates the days and the seasons and has helped to form mountains, plains, oceans and deserts. Like every deity it offers both threats and promises, and can condemn those who do not obey its rules to a painful death, but it does all this not with supernatural aid, but by obeying the banal rules of physics.

CHAPTER 2

KEEP OUT OF THE KITCHEN

Do not Bodies and Light act mutually upon one another; that is to say, Bodies upon Light in emitting, reflecting, refracting and inflecting it, and Light upon Bodies for heating them, and putting their parts into a vibrating motion wherein heat consists?

Isaac Newton, *Opticks* (1704)

The powers of sunshine give molluscs an unexpected role in theology. Lambeth Palace, the London home of the Archbishop of Canterbury, recognises their importance in its Resurrection Window; a post-war replacement for stained glass smashed at the time of the Reformation, reimagined in the nineteenth century, and destroyed once more in the Blitz. On the rocks below the risen Christ, and between the sleeping soldiers on either side of the empty tomb, are small, curled and yellowish objects, easily identifiable as snails.

They find a place because they, like the Saviour, appear to the untutored eye to die and to be born again. A species common around the Mediterranean seems to be lifeless in the summer months, but when the autumn rains arrive emerges miraculously from its shell – a behaviour seen by the early Church as a metaphor, if not more, for Christ's return after the Crucifixion. The Lambeth Palace window is true to the tenets both of its beliefs and of the landscape in which they find their origin.

But what of the science, rather than the symbolism, of the risen snails? In truth they find not rebirth but renewal. They have returned not from paradise but from a fiery underworld of the sun's making.

At some ingenious length, Psalm 58 warns the wicked that they will be destroyed in a variety of painful ways unless they accept the rule of God. It threatens sinners with heat-stroke: 'As a snail which melteth, let every one of them pass away'. For creatures that spend time on the ground, molluscs included, the dangers of a thermal inferno are all too real. To avoid that fate, some species climb up sticks to escape the heat of the surface. For several summer months the southern European landscape is then decorated with thousands of dried-out plants, their upper branches covered with small whitish shells. Their bearers can cope with ground temperatures of up to 40°C, but when the mercury rises above that, they abandon hope and haul themselves upwards, into cooler air. As the sun beats down they shut down body systems one by one and enter a state of suspended animation. Their internal machinery slows to a fifth of its active rate, and to save water they close off the mouth of the shell with a thick membrane. Then autumn arrives, the rains return, the air cools, and the animals are restored to life in their millions.

Even in Britain, on a summer's day the temperature near the ground can shoot up to lethal levels. A visit to an English sand dune on a day of sunshine and showers may reveal the fate of snails that drink too deep of the solar offering. Some have become trapped far from shelter as they search for food on the wet grass, and lie exhausted, their bodies extended, covered in slime – melting, in effect – and close to death.

The watching biologist can do little more than sympathise with their plight, for he or she will stay at a comfortable temperature even as the sun beats down. Large animals like ourselves take a long time to heat up or cool down, but even a shrew, which weighs little more than an adult garden snail, would manage with no difficulty in such circumstances, for it has, like all mammals, an internal thermostat denied to every mollusc.

In spite of that talent, men and women can also be caught out by the strength of sunlight. In the days when I worked on fruit-flies in Death Valley, a couple of our visits to distant oases were interrupted by blue flashing lights as the emergency services went to the aid of tourists overcome by the heat. There had over the years been several fatalities, most of them among drivers who had broken down on back roads and – in the days before mobile phones – set off on foot to find help. (Less often, a foolish visitor decided to walk across the salt flats to reach the snow-capped Panamint Mountains on the other side, for in the crystalline air they seem close, but are several waterless miles away.)

Heat has been a problem since life began. The earliest single-celled creatures were obliged to flirt with the sun and evolved mechanisms that allowed them to do so in safety. Their descendants, three and a half billion years later, from bacteria to blue whales, still share some of their tactics, which range from biochemical defence mechanisms to changes in behaviour.

Most animals live, like snails on a sand dune, on a thermal cliff. Much of their time is spent ensuring that they do not fall over it. Many creatures have to adjust their lives to the intensity of sunlight on a scale of minutes to hours, but although humans, like other mammals and like birds, can often afford to ignore its short-term demands, they too have to submit to them in a desert or on a glacier and as the seasons wear on. A failure to do so may be fatal.

Molluscs, flies, lizards, fish, frogs and their fellows are known as ectotherms, for their body temperature varies in accordance with that of their external environment. They tread a narrow path between accepting enough of the solar gift to warm up to go out and feed or have sex, and frying because they cannot reach the shade in time. Mammals and birds are not so often reminded of the risky world they live in, for they are endotherms, creatures blessed with the ability to control, within limits, their own internal temperature. The main difference between the two groups is not body temperature itself, but the way it is maintained. Birds and mammals are animated power stations fuelled by food and warmed by their waste energy, while fish, reptiles, frogs, insects and more soak up much of their body heat directly from the environment. Life for an endotherm is stable, comfortable and efficient, and allows a range of talents to flourish.

It is, on the other hand, expensive, for even at rest such creatures use twenty times as much fuel as does an ectotherm of the same size. Ectotherms teeter on a thermal tightrope more often than do endotherms, but both face the dangers of heatstroke or frostbite when conditions go beyond the safe limits.

For molluscs and men alike, the problem is at its worst on the ground surface. In Death Valley, summer air temperatures a metre or so above soil level can reach over 50°C (the record, measured in July 2013, was 54°C) which would soon kill. The ground itself is far hotter, for it stores a lot of the solar input that falls upon it and has almost no wind to cool it down. In Death Valley the surface can reach more than 90°C, and anyone foolish enough to expose themselves to that will not last more than a few moments. The Valley's original residents, the Shoshone Indians, knew that well and – rather like the Mediterranean snails – in the summer abandoned it for a cooler climate, up in the mountains.

A lot is known about the thermal relations of ectotherms such as lizards, snakes, butterflies and the like, in part because their movements in and out of the shade are easy to observe. Molluscs are less public about their tactics. I have spent large parts of my career, with mixed success, in attempts to persuade them to reveal how they do the job and how that might impact on their genes and their evolution.

The number of photons that such animals encounter depends on time and on season, on the shape, size and density of leaves and branches, on the number of gaps in the vegetation canopy, and on how much is reflected on to shaded areas by leaves lucky enough to pick up a sunbeam.

On a windy day a sun-fleck may last for seconds, on a still one for hours. Within an open woodland or hedgerow there can be more variation in temperature at ground level over a day than in air temperature over a month, and in hot places a thermometer placed on the surface can vary by 50°C within less than a metre on a day when air temperature varies by no more than 5°C. A snail anxious to stay safe has its work cut out as it tries to decide where to spend its time.

To watch those animals in the wild is tedious at best, because in dry weather they can disappear underground or hide beneath stones for days. I needed a way to assess their overall patterns of activity in daylight.

An idea of how to do so came on a trip to Cornwall in 1967. I remember it for a foul smell, a pivotal moment in popular music – and rather a good idea (the smell came from the oil that leaked from the wreck of the oil tanker *Torrey Canyon*, and the musical revolution began with the release of *Sgt Pepper*). One day, I had an 'only connect' moment in which I linked my interest in thermal ecology to the stability of dyes used in the plastics industry.

Up on the cliffs as we looked for sample sites we passed a board covered with electrical wires of various colours. I happened to mention this in the pub that night, and was told that its purpose was to test how much the pigments used would fade when exposed to the sun. In those days it was fashionable to wear blue jeans, most of all when they had begun to lose their colour because they had been exposed to ultraviolet. I at once had an eccentric idea: to cut out patches of denim and attach them to snails to see if the extent to which they faded could assess the time their bearers spent in sunlight. That did not work (and led to feeble jokes about jean manipulation) but another solution then emerged: to take the dye used in the cloth and mix it with a stable yellow paint. This gives a green compound that fades back to yellow when exposed to daylight. The method allowed us to ask questions about the relations of my favourite animals with sunlight, both in the wild in the Pyrenees and in the snail ranch of a hundred wire cages that we later set up at Oxford University's field station at Wytham Wood.

The ectotherms that feature on Lambeth Palace's Resurrection Window are a reminder that to clamber upwards, away from the surface, is an effective way to stay cool in hot weather. The green paint said more.

We found that in the Oxford cages, individuals of my own favourite species from Spain faded more than those from Scotland, because they climbed higher, no doubt because, in their homeland, even in cool weather, they face the constant danger of a sudden and lethal dose of the sun's rays if they stay too close to the ground.

When a snail feels too hot, it shows it. First, it flips up its head to give its brain a moment's relief, and a few moments later does the same to the tail. We tested that response on a warm hotplate. As the dial was turned up, degree by degree, they began their odd behaviour. Animals from Spain accept a much higher temperature before they start to flip than do those from Britain. Experiments with a drug that blocks pain receptors show that northern populations are more sensitive to heat-induced pain than are those from the south.

On sand dunes, and in Death Valley, what tends to kill is a sudden blast of heat rather than a longer period at a high, but more reasonable, temperature. For molluscs and motorists alike, what counts for survival is not the average of what they experience but the occasional crises when, perhaps for just a few moments, their internal machinery shoots over the lethal limit.

In what in retrospect proved to be a rather formative experience, both socially and scientifically, I spent several months after my A-levels and in subsequent university vacations working in a Merseyside power station (and I learned more about the laws of mechanics there than I ever did at school). One of my jobs was to check the valves in the pipes that carried superheated steam (the first move was to chip off the asbestos; we never wore masks, but fifty years on I seem to have got away with it). Each valve had to be kept below its safe thermal limit, and if it rose above that even for a few seconds it had to be replaced. The test used a translucent wax dot that covered a layer of black carbon dust and was placed on a sticky base. The waxes were mixtures of high- and low-melting-point compounds that could be set to melt from 40°C to 250°C. When the dot did melt, the dust rose to the surface. A valve that received the Black Spot was doomed.

We did the same with snails. We used the lowest-powered sensor available, and found that even in the moderate climate of Oxfordshire a few of the animals did reach 40°C on the shell surface, inflicting a heat stress which, if they did not escape in time, might prove lethal. Further south in

Europe the proportion facing such a stress must be considerably higher.

All this is straightforward thermal ecology, but my main interest has always been the interaction of that science with genetics. Our work in Croatia and across Europe had shown that temperature changes on a scale from a few metres to thousands of kilometres alter the incidence of light- and dark-coloured shells. But why, then, within almost every patch of dune, hedgerow or grassland, do populations retain a variety of genes for shell characters? Why are some dark individuals present at the southern limit of the species' range and why, on the chilly coasts of Scotland, are there a few light-coloured ones? Perhaps, we thought, thermal differences on a scale of centimetres are enough to do the job.

Because of its importance in crop production, a whole industry measures the penetration of sunlight into vegetation. It uses fish-eye cameras to take pictures through the leaf canopy to measure how much sky is exposed, or thermal sensors that take thousands of measurements within a square metre. A long-wave radar based on laser pulses from aircraft can even draw a three-dimensional picture of the structure of the leaves and branches. These methods are ingenious, but given the modest support available to those who study molluscs, we needed a cheaper way to get a snail's-eye look at the sun. Once again, the fading green paint, this time with the help of a device popular in Japanese game arcades, came to the rescue.

Take ten polystyrene balls the size of snail shells, mark off a square metre of bushes, nettles or grass, and throw them in at random. As in a pachinko machine, the balls follow a path that depends on the number of obstacles in the way. On short grass, almost all reach the ground and are exposed to the open sky. Replace them with steel discs painted with light-sensitive paint and wire them flat into the vegetation just where each ball comes to rest. After a week, pick up the sensors, and measure how much they have faded from green to yellow.

We did the experiment in the Spanish Pyrenees, where snails are abundant from sea level to around two thousand metres. Genes for light shell colour were – as expected – more abundant in places with most sunlight, and all the animals on alpine pastures above the treeline were pale yellow with no dark stripes. Even better, shell diversity in each place was tied to the degree of variation in exposure to its rays. A dappled environment supports a more variable population.

Perhaps, we then thought, such patches of sun and shade enable genes for dark and light shell to coexist because each type occupies the part of the habitat in which it feels most comfortable. We marked thousands of animals with small spots of the paint and put them out in the Pyrenees or in the Oxfordshire cages. In a few weeks a marked difference between dark and light individuals emerged, as a signal that they had indeed been exposed to the sun for different periods, either because they chose to live in different parts of their habitat, or because they were active for longer or shorter periods in daylight. The shell variation may hence allow the population as a whole to occupy a wider range of thermal habitats than if all were dark, or all light.

That discovery may be of interest to evolutionists, but has had little impact on biology as a whole. The problem is that snails are for gourmets, but those who prefer the bread and butter of genetics are more interested in fruit-flies. Those creatures were at the centre of that science a century ago, for they were used in the first experiments to map genes on to chromosomes, to detect mutations, to study development, and more. After a brief dip in popularity they have today, thanks to technology, regained some of their prominence. Land molluscs may be in their twilight years, but fruit-flies – although they must now defer to ourselves for star billing – have seen a new dawn. We decided to join the chorus.

The first cellular response to heat was discovered more than a century ago when, in an early sign that genes switch on and off in response to an external stress, it was noticed that a sudden blast of high temperature causes the chromosomes of mosquito larvae to swell up at specific points. Five decades later, in experiments on fruit-flies the specialised proteins produced – the heat-shock proteins, as they are called – were found. Many among them survive almost unchanged from our most remote ancestors. Most creatures have dozens of such things, involved in responses to pain, in ageing, and in development. In mammals they also play a part in the immune system and in cancer. Arctic fish summon them up at just 5°C, while bacteria from hot springs do not bother until the liquid has almost begun to boil. In the same way, snails from cool places switch them on at a lower temperature than do those from hot. Some are not generated until the outside agent arrives, and take hours or days to swing into action, but others lie alert like a cocked pistol, ready to respond within seconds. A burnt finger summons them up at once, but a move from the Scottish to the Samoan climate wakes them up in a more gradual fashion, and over a few weeks helps the body to acclimatise. As a result it takes less to inflame the inhabitants of Novosibirsk than it does those of Naples, for the Siberians summon up their proteins at lower temperatures than do the Italians.

These talented proteins – now, in response to their role as guardians against a wide range of stresses as well as heat, rather charmingly referred to as chaperonases – unite the whole of life in a shared reaction to a hostile world and hint that a battle with the sun has moulded existence since it began.

In spite of the importance of Drosophila in research on heat stress, little is known about its own thermal relations in the wild. In a brief excursion into that question we used the flies themselves as animated thermometers.

The eyes of the most widely studied species, *Drosophila melanogaster*, are usually red, but mutations can change the colour to white, brown, apricot and other hues. One among them alters the rate at which pigment is made in a way that depends on temperature. In a fly that develops in cool conditions the eye is dark brown, but when the animals grow up in a warmer place they find it hard to make the pigment and the structure is almost white. A glance at the eye of any mutant adult is hence a hint of the temperature at which it spent its early life. The crucial moment comes when it is laid down in the pupa, the stage between larva and adult.

We released around half a million flies bearing this mutation in two sites, one in the lowland experimental station of the United States Agricultural Station in Beltsville, Maryland, north of Washington DC (our location was then called Pesticide Road but has since been renamed, for reasons of political expediency, Biocontrol Road), and the other in the Blue Ridge Mountains, eighty kilometres away and a thousand metres above sea level. Each site was provided with a large pile of rotting fruit, on which the experimental flies laid eggs. The fruit was in the sun for much of the day, and its surface temperature rose to 44°C at times, while on the ground below it was 13°C cooler.

Their offspring's eyes were checked as they emerged and were trapped over subsequent weeks. They showed that the pupae had developed in temperatures that ranged from 15°C to as much as 29°C, which is within the range at which the animals improve their heat resistance because protective proteins are switched on. In addition, flies raised in the laboratory at those temperature extremes have large differences in their rate of development, their size, their sexual success, and their survival.

As had the Spanish and Scottish snails, each population changed its behaviour to suit the local climate. The lowland and the mountain release sites had an average temperature difference of just under 3°C, but the difference experienced by the flies themselves averaged a little over a degree, as a hint that members of each population had adjusted their position in the bait to occupy the optimal habitat. Again as in the snails, there was plenty of individual variation in exposure to heat and cold, for the temperature experienced by individual insects varied from 16°C to 29°C in the lowlands, and 15°C to 27°C in the mountains. Fruit-flies, like snails and sunbathers, adjust their relationships with the energy that streams in to stay within a safe range but, even so, experience individual differences that must have large effects on their survival and reproductive success.

To work on the thermal lives of invertebrates is agreeable enough, but at first sight seems to have little relevance to our own lives. Most of the time, we have a more relaxed association with the thermometer than do molluscs or insects, but without our inner thermostats and furnaces, our lives would be as uncertain as is that of a snail on a sunny day. When, where and why did that mysterious mechanism emerge?

Snails, flies and their fellows are often referred to as cold-blooded, in contrast to the warm-blooded mammals and birds, but those terms are too simple. The temperature of a Mediterranean snail on a sunny day may rise above our own, while bats that hibernate in an icy cave can stay almost at zero for weeks. Even that pales when compared with the edible dormouse, which falls into torpor with a dramatic drop in body temperature for as much as ten months a year (it lives, not by coincidence, for far longer than most mammals of its size). In addition, animals often seen as cold-blooded generate quite a lot of internal heat. A tuna spends much of its time near the warm surface, but to feed must plunge into the depths, where life is much colder. Heat generated by its massive muscles is shunted to the head with a system in which warm blood on the way back from that part of the body passes its stored heat to cooler blood on the way in, and keeps the brain at a higher temperature than the rest of the body. In much the same way, pythons as they incubate their eggs shiver to push up their internal thermometer. Even plants, those servants of the sun, can warm themselves up, for a shoot of North American skunk cabbage can push up its own temperature to an extent sufficient to allow it to melt its way through spring snow.

The coolest mammals live in the warmest places. They include the platypus and its relative the echidna, together with South American sloths and anteaters. Carnivores – dogs, lions and the like – have hot bodies, with weasels warmest of all. Because flight burns so much fuel, birds are, on average, several degrees hotter than mammals, with penguins and ostriches at the bottom of the scale, and sparrows and their kin at the top.

In all those creatures, the machinery of the body – digestion, circulation, cogitation and exercise – generates some of the internal heat, but when life gets cold they summon up other mechanisms. Shivering can boost heat production by five times, but because it uses so much oxygen cannot be sustained for long. Some mammals, and all birds, do so in a way not visible to the eye, for they use individual groups of muscle fibres rather than whole muscles. When the temperature drops, muscles can switch from physical work, with warmth as a by-product, to the generation of heat as their main job. The 'brown fat' found in mice and in human infants acts as a reserve that can be burned in an emergency or even on a cool day. About a sixth of the body weight of a new-born baby is made of fat, more than in any other mammal, perhaps to insulate a naked and vulnerable child.

Endothermy opened the door to a host of opportunities. Warm blood makes for busy bodies and great minds. Many more mammals survive as herbivores than do lizards or frogs, perhaps because a high-temperature gut makes it easier to digest plants. They soak up more oxygen and can run further, and faster, than can any endotherm of the same size. Even more important, they have evolved a large, and expensive, information system, the brain.

A lizard can move fast when it has to, but is tired out within a few minutes and may take hours to recover. Like a runner as he sprints towards the finish, the animal uses a form of muscle activity that does not use oxygen and generates lots of toxic by-products. Most of us never experience the cramps suffered by lizards and top sportsmen, because our efficient lungs and vigorous hearts soak up all the oxygen we need.

The road to inner warmth began in the days of the dinosaurs. Like today's lizards, they basked in sunlight to warm up. Once such creatures had reached a critical size, their bodies were large enough to begin to store heat in their core, and they soon evolved the ability to shunt the flow of blood through vessels that would keep it there, or to lose it from the skin, according to demand. Today's Komodo Dragon – a large Indonesian lizard – can in this way keep its internal temperature at about 6°C warmer than the air around it, which takes it to just a few degrees below our own.

The largest dinosaurs weighed fifty times more than that creature and may have been able to maintain an internal temperature close to our own. All mammals have specialised bone structures within the nose that retain heat in the body as they breathe in and out. Some dinosaurs had such heat-exchangers long before mammals evolved, which suggests that they too were in control of their own body heat. A few became energetic beasts, able to stalk and kill their cold-blooded relatives, while others found a home in northern Alaska, which was warmer than today, but still had short winter days with frosty nights. A warm body may also have meant that some dinosaurs could, like pythons, incubate their eggs. That enabled them to hatch earlier and grow faster, and perhaps helped their descendants, the birds, to take their great leap into the air.

The first dinosaur to sport feathers emerged almost two hundred million years ago, and those structures, too, were first used to keep the animal warm. The smallest among them, the proto-birds, took flight some fifty million years later, and in the sky's empty arena diversified fast into the many kinds we know today. The first mammals, too, had an explosion of evolution with the early emergence of the groups that gave rise to mice, mammoths, bats and whales. To take control of one's own temperature was a milestone in the history of life.

There was, however, an interlude in which those mammal-like ancestors seemed to fade from view. Their descendants – ourselves included – have a lot to be proud of, but compared with lizards and their kin they lack vision.

Only their eyes feed in information about the sunlit world, while other creatures can sense light that falls on other parts of the body such as the skull. In addition, mammals have at most three colour receptors, while many birds, fish and reptiles have four. Scent, hearing and touch are all in contrast more developed in ourselves and our relatives, while their eyes are unique among the vertebrates in that they have rod receptors, which specialise in dim light. All this hints that in the days of the dinosaurs our ancient ancestors moved underground and came out only at night (a lifestyle impossible for an ectotherm).

Useful as warm bodies might be when avoiding dinosaurs, they cause problems of their own. Every furnace makes waste heat. Human muscles, too, are inefficient, for three-quarters of their effort produces heat rather than work. Without an ability to cool itself, the internal machinery would cause the temperature to rise at a degree an hour, and death would not be long delayed. A trained athlete performs best at an air temperature of 11°C, but his capacity drops fast as the thermometer rises above 21°C (not unusual in a British summer). Ten degrees higher still, and all exercise becomes dangerous as the body is pushed beyond its ability to lose heat faster than it generates it. On a warm summer afternoon, a marathon runner, or even a hod-carrier, is at real risk of collapse. In some places and seasons millions of people spend a good part of their lives almost as close to a lethal temperature as does a snail on a sunny day.

The vulnerability of mammalian flesh to even moderate temperatures was first noticed in an eighteenth-century kitchen. The observation was used to develop a radical new culinary technique that illustrates the body's impressive ability to cool itself down.

A few years ago, a famous – and famously bad-tempered – television chef opened a grand restaurant near my home in Camden Town. With the meanness of spirit typical of the British press, one tabloid accused him of having precooked food delivered, rather than making it himself. That claim was unjustified, as the indignant proprietor pointed out, for he prepared some of his meat and fish dishes 'sous vide', a technique in which raw materials were sealed into plastic bags in another of his premises, and then, in the restaurant itself, placed in a heated bath for an hour or so. That produces a tasty piece of steak or salmon.

The method is often assumed to be modern, but in truth is not, for it goes back to the eighteenth century. It was invented by the British physicist Benjamin Thomson, Count Rumford (who later helped found the Royal Institution). He was interested in the use of science in the home. Rumford was scornful of the stoves, fireplaces and oil lamps of his time, and did much to improve them (he also invented thermal underwear and the pressure cooker). One of his many experiments seemed at first to have gone wrong but – as so often in science – instead produced an unexpected result, with important consequences:

> Desirous of finding out whether it would be possible to roast meat in a machine I had contrived for drying potatoes, I put a shoulder of mutton into it, and after attending

> to the experiment three hours, and finding it showed no signs of being done, I concluded that the heat was not sufficiently intense; and despairing of success, I went home, rather out of humour at my ill success, and abandoned my shoulder of mutton to the cook maids . . . When they came in the morning to take it away, intending to cook it for their dinner, they were much surprised to find it already cooked, and not merely eatable, but perfectly done, and most singularly well-tasted. This appeared to them the more miraculous, as the fire under the machine was gone quite out before they left the kitchen in the evening.

The idea of slow cookery at low temperatures, the precursor of sous vide, was born.

In its modern guise, a steak can be parboiled to medium-rare at no more than 54°C. That raises an interesting question. People in saunas often experience air temperatures much higher than that, sometimes even above boiling point. Why, made of meat as they are, do they not cook?

The answer can be found in an alternative form of purification, the Turkish bath. There, the maximum tolerable temperature is no more than 48°C. The crucial difference is that a sauna works with dry heat, while in the Ottoman version the air is saturated with water vapour. Our cooling machinery – the evaporation of sweat most of all – cannot work in such conditions, and the body pays the price.

The rate of the heat loss from the body core to the surface varies by around six times as blood vessels contract or expand. From there the excess is shed by conduction, by convection, and by radiation. The body also uses evaporation, with the help of its three million sweat glands.

The role of sweat was identified in the late eighteenth century by the physician Charles Blagden. In a report to the Royal Society he tells of an experiment in which a dog was placed in a room heated to 113°C, together with some eggs and a steak. Within a quarter of an hour the foodstuffs were cooked, but 'The dog panted and held out its tongue, but the symptoms did not show evidence of ever becoming more violent than they are usually observed in dogs after exercise in hot weather'. Its body temperature was almost normal but 'we opened the basket and found the bottom of it very wet with saliva' as a signal that it could cool itself by panting. He then put himself through the same experience:

> The first impression of the heated air on my naked body was much more disagreeable than I had ever felt it through my cloaths; but in five or six minutes a profuse sweat broke out, which gave me instant relief, and took off all the extraordinary uneasiness . . . it appears beyond all doubt, that the living powers were very much assisted by the perspiration; that cooling evaporation which is a further provision of nature for enabling animals to support great heats . . . whenever I tried the heat of my body, the thermometer always came very nearly to the same point.

As the dog panted and the scientist mopped his brow, evaporation saved each of them from being boiled alive.

Two centuries later another researcher built on Blagden's heroism. He ran as hard as he could on a treadmill as the temperature was raised from 45°C below zero to 55°C above. Over that range – almost enough to turn ice into steam – his own temperature varied by less than a degree. At the start he had to use a large amount of effort to get warm, and at the end had to utilise even more to avoid heatstroke. The test had to be stopped when his blood pressure shot up. Whatever the risks, his ordeal gave eloquent witness to an endotherm's ability to withstand the extremes.

Every mammal stays cool in its own way. As Blagden had noticed, dogs pant when they are hot – at up to four hundred times a minute. They have no sweat glands, but their lungs are so elastic that it does not cost them much to do so (human ribs are much stiffer and panting generates more heat than it loses). Some primates urinate on themselves to stay comfortable, while rats and mice smear saliva over their fur to do the same. We have less need for such eccentric habits, for sweating is, at least up to a point, an efficient way of cooling down. However, it does have its dangers. A trained athlete on a hot dry day can lose three litres of water an hour in sweat, but the gut cannot absorb more than about two-thirds as much, so that even if he or she drinks enough there remains a real threat of overheating. The unfit, the old, the young and the obese are all at greater risk, and they may be in danger at temperatures well within the range of those of a Mediterranean, or even a British, summer.

Humid heat is much more hazardous than dry, as the body cannot then cool with the help of evaporation. A measure that combines temperature with humidity – the wet-bulb temperature – is used to give an estimate of how hot it actually feels, and reveals why a thirty-degree day in steamy Washington seems so much less comfortable than does the same temperature in arid Los Angeles. The device is simple enough, and is not used as much as it might be. It consists of a regular thermometer whose bulb is wrapped in a cloth soaked in water. As that evaporates, it brings down the mercury, but on a day of high humidity it cannot do the job and the figure creeps up.

A wet-bulb temperature of 35°C will kill even a young and fit person at rest in the shade within a few hours. In July 2015 in the southern Iranian coastal city of Bandar Mahshahr that level was almost achieved. Thirty-two degrees may prove fatal to an older individual, or someone not in the best of health, and that was reached in Sydney's Western Suburbs two years later. For a labourer or peasant at work on the roads or the fields, the absolute safe limit is seen as just one degree lower than that. Places in India, Pakistan and China often experience a wet-bulb level of 28°C and sometimes nudge 30°C, which is intensely uncomfortable for everyone and very dangerous for some.

The effects of heatstroke are unpleasant at best. First comes heat exhaustion, as sweat runs off the body. As its temperature rises above 40°C, the breath becomes laboured and the heart beats faster as it strives to move blood to the surface. Those afflicted may fall into a coma, or become delirious. Arteries and veins then relax and blood pressure plunges. As the system abandons hope, death comes through brain damage, muscle breakdown, kidney failure and general collapse.

Plenty of people have suffered that fate. In France, the *canicule* of August 2003 killed, according to some estimates, as many as forty thousand. The dry-bulb thermometer sometimes rose to 44°C, more than ten degrees above normal, and dropped by just a fraction at night. Too few funerals could be organised, and some corpses had to be kept for weeks before they were buried. In Russia seven years later fifty thousand died of the same cause.

In London, mortality rates change little with temperatures between 10°C and 24°C, but above that the figures begin to climb, with a three per cent increase for every extra degree, so that at 30°C – a level now reached several times a decade – the death rate goes up by half. Not all is gloom, for society now copes much better with such events. In the United States, and even more in Australia, there has been a decrease of seven-tenths in the numbers of deaths from heatstroke from the end of the nineteenth century to today, mostly because of the replacement of men with machines on fields, roads and construction sites. In the successive heatwaves of the present century the more recent has always killed fewer people than its predecessor. The 2003 European event was followed by another a decade later. The number of deaths in the latter went down by four-fifths, in part through new public awareness of the dangers, through improved weather forecasts, the availability of better medical treatment and, no doubt, the spread of air-conditioners. Even so, to blunt the good news, an ageing population means that more of us are now at risk.

Plenty of people die of heat shock even on cool days because of their own foolish behaviour. Some drugs crank up the body machinery to lethal level. Users like to be stimulated, but sometimes the stimulus goes too far.

Methamphetamine ('crystal meth') leads to euphoria but can also spark off a sudden rise in brain temperature that can kill, or – perhaps worse – lead to an instant onset of severe Parkinson's disease. Cocaine, too, may throw skeletal muscles into cycles of contraction and relaxation. This generates lots of heat, but because of the intense activity the arteries cannot provide enough blood to cool them down. The muscle cells release their contents, which damage the kidneys, often to fatal effect. Ecstasy, popular among British students, leads to elation, frantic dancing, a sense of communion with one's fellows, and too often a lethal collapse. For those struck by such disasters the best treatment is to plunge at once into an ice bath.

The body's thermostat, when not pushed beyond its limits, is also an important part of its defences. The signs of fever were known to the ancients as *calor* (heat), *rubor* (redness), *tumor* (swelling) and *dolor* (pain). Mammals, birds, reptiles and even insects all push up their temperature after an infection, so that the process must have evolved long ago. It costs a lot, and for men and women a rise of just 1°C involves an increase of about a tenth in metabolic rate.

The notion that fever is a defence rather than a disorder was first put forward by the seventeenth-century physician Thomas Sydenham, who referred to it as 'Nature's engine which she brings into the field to remove her enemy'. He was right.

There are strict limits as to what is safe, and a body temperature that rises above 38.5°C is a threat. Even so, most fevers are not an illness in themselves but a statement that the system has started to fight back. A high fever may reduce virus replication by a hundred times, and can work almost as well against bacteria (although some among them retaliate when they switch on their own protective proteins). Feverish patients given drugs to cool their brows do not improve their rate of recovery and may indeed be at increased risk. Even to treat flu in this way can kill.

In response to an infection, the hypothalamus, at the base of the brain, turns up the internal thermostat. The 'set point' – the level at which its owner feels comfortable – is pushed upwards, which is why, even in a warm room, feverish patients complain that they have a chill. Often they shiver, which, although it may persuade them that they do indeed feel cold, generates extra heat. A surge of adrenalin pushes up metabolic rate and conserves heat in the body core. It also alerts the immune system, which swings into action.

Hippocrates tells of an attack of malaria that much improved the symptoms of a patient with epilepsy. Much later, the German psychiatrist (and in later life ardent Nazi) Julius Wagner-Jauregg noted that a woman in the midst of a psychotic episode recovered when she had a severe fever brought on by a skin infection. He then had a bright, albeit somewhat unorthodox, idea. During the late stages of the First World War a German soldier with malaria came to his clinic. Many of its psychiatric patients were in the last stages of syphilis, which leads to profound mental disturbance. The physician took a sample of the soldier's blood and injected it into several of them. Soon, all developed a high temperature. Around half showed real improvement in their mental problems, and some even returned to normal life (about one in six did die of the tropical disease, but as he pointed out to their grief-stricken relatives, they would soon have perished of their venereal infection anyway). Wagner-Jauregg was honoured with a Nobel Prize for his research, and in time, thousands received this treatment, with considerable success. We now know that protective proteins are summoned up during a fever and that may have been responsible for their improved prospects. Infection with malaria parasites was succeeded by a system in which patients were placed in a heated room or a hot bath. That too worked quite well. With the discovery of penicillin, thermal treatments were abandoned. In these days of antibiotic resistance they might find new life.

Fever has also been used to treat cancer, with even Louis Pasteur in favour of the idea. Some physicians once went so far as to apply pus-soaked bandages, or later a cocktail of bacterial toxins, to the open wounds of advanced cases to push up their thermostats. Experts now dismiss that approach, but some patients do respond to heat treatment, either in the affected organ or in the whole body.

Ectotherms, too, respond to illness with a rise in body temperature, but they do it in their own way. Some lizards that spend most of their time in the shade bask in the sun if they pick up a bacterial infection even while this increases the chance of being eaten by predators, for if they are kept away from its rays their mortality rate doubles. As a reminder of our own ectothermic past, people in the first stage of a cold or flu may sit in front of a fire, or go out into the sunshine, perhaps as part of a temperature-based defence.

One ancient shift in human thermal behaviour sets us apart from all our relatives, for it put mankind on the road to what he has become. Like snails, our ancestors faced the need to escape from the heat on the soil surface. The attempt to do so led to a new talent, the ability to stand on two legs.

Chimpanzees spend much of their time in the trees, and at ground level can do little more than lurch along on all fours, or run for a few bipedal paces. Around seven million years ago, our own ancestors began to come down to earth on a more permanent basis. They did so as mountain ranges grew in eastern Africa and cut off the Rift Valley, a centre of human evolution, from the wet winds that once blew in. The skies cleared and the landscape opened up. In time, the terrestrial primates – who still found it hard to stay upright for long – were forced into open country, where they were battered by the sun.

Under the cloudless skies of Africa, the temperature on the ground can reach intolerable levels. The permanent crouch that had worked well in the shade took our predecessors too close to that inferno. To find cooler air, they did not climb back into the trees, but stood upright. At head height, air temperature reduces to more tolerable levels, often with the help of a fresh breeze, and, as a bonus, much less of the body surface is exposed to radiation than when on all fours. Quite soon, our ancestors learned, or evolved, to stand on their own two feet.

The vertical habit opened the door to a new way of life. Bipeds could travel further and faster, and became better hunters. As groups grew larger people were forced to deal with social stress. Their brains grew to match and, in time, a modern society emerged. We owe more of what makes us what we are to the power of the heavens than perhaps we realise.

It did not take long for our grey matter to begin to challenge their rule. The earliest evidence for man-made fire is about a million years before the present. At first the flames were used just for cooking, but in time they provided a sociable place to sit around on a cold East African night.

Clothes came later. Scrapers that might have been used to clean hides for garments are found from seven hundred thousand years ago, but they could have been used to cut meat instead. Another clue hints that we stayed naked for rather longer than that.

Humans have head lice and body lice. The latter must have emerged from the former, for it lays its eggs on clothes rather than hair. Assuming that mutations accumulate at a steady rate, the genetic differences between them can be used to estimate when the split took place. They hint that the body was invaded from the head around a hundred and fifty thousand years ago, perhaps with the invention of the first garments.

Caves kept people warm well before that, but earliest artificial shelters are rather recent, for their remains do not appear until the latter part of the Stone Age, ten to fifteen thousand years before the present.

Fire, clothes and houses much improved mankind's ability to cope with a shortage of sunbeams. They allow men and women to reduce the constant adjustments needed to cope with daily or seasonal changes in temperature and have enabled us to spread from the tropics almost to the poles.

From the Stone Age to the mid-twentieth century, although fires became more efficient, clothes warmer and houses more weathertight, not much changed apart from the details. Since then, in not much more than five decades, our thermal strategies have been transformed. The revolution began in the United States but has spread across the developed world. We are now more sheltered from the challenges posed by the weather than at any time in history.

Once, men and women in cold places did not need to do much more than light the fire, retreat into a hut and wrap themselves in furs to stay comfortable, while those in the tropics found life even easier, for they had only to seek the shade when necessary. Winter life in my icy Edinburgh basement was not much different from that in Boswell's day. I spent that season wrapped in string vests and thick sweaters. In my room, I fed sixpences into a gas meter, and at night, in those days before duvets, I piled on the blankets.

Now, a whole new way of life has emerged. Once, people coped with heat and cold by warming, or cooling, their bodies, or a small part of the room they lived in, but now, from the tropics to the poles, we control the temperature of our entire environment, from houses to offices and from supermarkets to cars.

Homo sapiens has become, in thermal terms, the most

profligate animal ever to have lived. Much of the output of the world's power stations now goes to maintain the optimal temperature indoors – around 23°C for most of us, with little difference between tropical peoples and those from temperate climes – whatever the conditions outside. We do it not with sweaters and blankets, T-shirts and shorts, or with gas fires and electric fans that produce patches of warmth or cold, but instead by regulating the whole climate in which we spend our time.

Today's cities have a metabolism of their own. It ensures that their inhabitants almost never need to call on their biology, or their wardrobes, to stay comfortable. In this new world, it has become as easy to watch television in underpants in a Stockholm January as to wear a heavy suit in the office in a Washington August. The inhabitants of each place experience the same climate as long as they stay indoors (which means, for most of them, most of the time).

In half a dozen medium-sized North American cities, from Timmins, Ontario, four hundred kilometres north of Toronto, to Key West at the southern tip of Florida, fuel bills from thousands of homes reveal the cost of that new way of life. In Ontario the average January low is minus 23°C and the summer high 24°C. In Key West, the figures are 18°C and 32°C, with a thermal range far narrower than that of its northern cousin.

Before Columbus, the peoples of the Americas, from the Caribbean to the Arctic, shivered or sweated, sheltered or sunbathed, spent more, or less, time in the sun, and stayed almost naked or wrapped themselves in furs. In subtropical Key West, it would have been possible to live outside unclothed all year with no artificial aids at all. That was not, needless to say, the case in Canada, but human ingenuity meant that its inhabitants could survive without much difficulty; after all, the Inuit of Alaska, with their caribou pelts and sealskins, their igloos and their seal-oil lamps, coped with temperatures of minus 50°C, far below what the Ontarians had to face. Now, at the cost of lots of carbon-based fuel, citizens of both places share more or less the same climate.

The new thermal world began in the United States. Even two centuries ago, British visitors found the heat indoors in winter intolerable: Charles Dickens wrote of his 1842 visit to Boston that his hotel was made 'so infernally hot (I use that expression advisedly) by means of a furnace with pipes running through the passages that we can hardly bear it'. Almost all the inhabited spaces in the colder parts of the country now have central heating, often to the temperatures that so discommoded Dickens.

In summertime, air-conditioning has almost become a human right. Even in the 1930s tenement dwellers in New York would spend whole nights on the roof to escape the August heat, a practice now more or less extinct. The first air-conditioner appeared in a Brooklyn print-works in 1902. At first, such devices were used to improve labour productivity, but they soon invaded the household. Now nine out of every ten American homes and an even higher proportion of offices have turned into habitable refrigerators (and half a century ago just one in ten cars was so equipped, while today almost all are). As an incidental, the invention led to the rise of shopping malls, office skyscrapers, and cinemas that stayed open in summer. The heat extracted has to go somewhere, so it goes outside, to make life even more miserable for those who cannot afford to frequent such places. On a larger scale, the air-conditioners also sparked off a mass migration to the Sunbelt, which now holds a far higher proportion of the American population than it did in the 1950s. Today, such machines consume about one-sixth of the total energy used in the United States – which represents more electricity than the whole of Africa needs to satisfy all its demands, from street lights to heavy industry. The country's new approach to climate already uses twenty times more than did the Native Americans, and is far more profligate than that of the Pilgrim Fathers or the inhabitants of Boswell's or my own Edinburgh.

These devices have become almost universal. The Chinese buy around fifty million a year, while more than half the entire output of electricity in Saudi Arabia in the summer is used to keep houses and offices at an acceptable temperature.

All this comes at a price. If we cannot stand the heat, we can still choose to keep out of the kitchen, but more and more of our fellow citizens will decide to flip a switch instead. As they cool down, or heat up, their kitchens, their living rooms, their workplaces and their cities, the world outside will warm in response. The consequences may in time force us all back into a way of life closer to that of the nineteenth century than of the early twenty-first.

CHAPTER 3

AN OCEAN OF ANOMALIES

All the waters run to the sea and yet the sea is not full, and from the place where they began, thither they return again.

Ecclesiastes 1:7

Man's dependence on his local star is so obvious that we sometimes fail to pay due attention to its broker here on Earth. The chemical that does the job, dihydrogen oxide, H_2O – water – may seem rather an ordinary fluid, but it is not. Among its many peculiarities, water is the only substance on the planet to be present simultaneously as a solid, a liquid and a gas. That attribute enables it to spread the gifts of the sun from the deeps to the edge of the atmosphere, to make the weather and the landscape, to allow plants to soak up the vital fluid from the soil, and to lubricate life itself.

Such talents have given the liquid a central place in literature, in art and religion, with *Moby-Dick*, Hokusai's *Great Wave* and more. The Bible has seven hundred references to the stuff. Greek mythology, too, was a soggy place, with an underworld surrounded by rivers of woe, oblivion and other doleful sentiments. Immersion in the Ganges is, in contrast, said to ensure eternal health. That may be true in spiritual terms, but in corporeal emphatically is not. Each day the river swallows up three billion litres of untreated sewage, together with the ashes of hundreds of corpses burned on its banks (to which are added in their native state the remains of holy men, suicides, pregnant women, young children, lepers, victims of venomous snakes, and those whose families are too poor to afford the firewood). In the auspicious year of 2013, the river's Kumbh Mela festival, held once every twelve years, attracted thirty million people on a single day in the largest crowds the world has ever seen. At such times, its waters contain levels of antibiotics (freely available as they are in India) equivalent to those in the bloodstream of patients under treatment for infection in Western hospitals. On the sacred banks of the nation's symbolic mother, myth and reality – as so often – make uneasy bedfellows.

Thales of Miletus, who lived in the seventh century BC, is sometimes hailed as the first prophet of the advantages of rationalism over such blind belief. He was the earliest philosopher to come up with a natural hypothesis of how the world began. Everything, he wrote, is water. His statement was perhaps a little overstated but has a kernel of truth. The substance plays a large part in the Earth's physics and chemistry, and without it biology – and those who study it – would not exist.

The elixir that fills our planet's veins was born in the

stars. In its earliest days the sun poured out vast quantities of hydrogen and oxygen. The two elements linked up on grains of stellar dust, and coated them with ice. The dust made clouds that fused into boulders and at last into planets, our own included. At first our homeland was hot, with a steamy atmosphere, but as it cooled it rained, in a tempest that lasted for millions of years.

Our cosmic neighbours had much the same history, but squandered their endowment. Mars was once a showery place. Many of its rocks form only in the fluid's presence, and so much is still trapped within them that if it were released the planet would be swamped. Its surface is marked with the relics of lakes, rivers and glaciers, together with those of seas as large as the Arctic Ocean. Ice at its poles waxes and wanes with the seasons, and in a few places near the equator, for a few days each Martian summer, brine flows downhill before it freezes again. At a latitude equivalent to that of Edinburgh, the Red Planet has cliffs of ice a hundred metres high. Much more is hidden below in a salty and liquid sea, twenty kilometres down. A century and a half ago came the mistaken claim that its landscape was covered in artificial canals that were extended from one year to the next. The idea was dismissed by Alfred Russel Wallace in his last book, *Is Mars Habitable?*, which made the case that the Martian navvies were a fantasy. Inhabited or not, the Red Planet was at least once wet.

Mars dried up because it was too small to hold on to its precious fluid. Its atmospheric pressure is one hundredth that of Earth, and most of its water evaporated long ago. Venus is too hot, with a dense atmosphere of carbon dioxide. Once, it had oceans. They disappeared when the greenhouse effect made it the hottest planet of all, with a surface temperature of around four hundred degrees. Even worse, Venus lacks a shield against solar radiation, so that what little remained was broken down by those rays. The planet still emits a trail of hydrogen and oxygen that streams away in the solar wind. Earth is in the sphere of Just Right, and has kept its precious legacy and all that flows from it.

Jules Verne speculated in his 1864 novel *Journey to the Centre of the Earth* that the bowels of our own planet might contain forgotten oceans ready to be explored. With the benefit of hindsight he was right, for the Blue Planet is as blue on the inside as on the out. Shifts in the ratios of dissolved gases show that a lake two and a half kilometres down in a Canadian copper mine has lain undisturbed for two and a half billion years. Thirteen kilometres deeper than that, beneath an Andean volcano, lie half a million cubic kilometres of water squeezed at high pressure into the pores of rocks. Yet more is hidden at even greater depths, contained in a blue-green compound of magnesium and silicon. Altogether there may be more water underground than in all seas, lakes, rivers and icecaps combined. Much of it dates from the planet's birth.

My physics teacher was right when he informed me that it only rains when there are clouds in the sky. That simple observation is a clue to the unique abilities of water. Without its endless cycle from solid to liquid to vapour and back our planet would be as dull and dusty as he was.

Where the rains come from was known to the people of India a thousand years before Europeans had any idea. The Upanishads, written there from about 800 BC, gave an accurate account of the tie between clouds and rain. Much later, Aristotle insisted instead that air was turned into water in deep caverns, while his successors thought that the seas were distilled by the planet's internal heat and the vapour blew onto the land. Even Leonardo da Vinci believed that rainfall was not enough to feed the rivers, and that most of their contents came from reserves beneath the oceans. Not until 1674 did the Frenchman Pierre Perrault get the story right after he had measured the flow of streams, the absorption of rain by the soil, and its rate of evaporation. He had discovered the water cycle.

To the people who live in it, the aqueous world seems full of movement. Rain, snow, waves, tides, rivers, streams, waterfalls and glaciers – all speak of change. Our lives, too, depend on a constant traffic of the stuff within our bodies and across the world. Now its true restlessness has been revealed. A network of satellites, of balloons, telescopes and weather stations, and of floats set loose in the oceans to plunge to the depths and return to the surface, tracks its travels on every scale.

The ocean contains ninety-seven per cent of the Earth's surface water and is by far the biggest player in the endless jousts between its three phases. Just one part in fifty of the H_2O is locked up as ice, even less lives in lakes and streams, and no more than one molecule in a hundred thousand makes an excursion into the air. Streams hold on to their contents for a few days, while a large river does so for around six months. In deep lakes, the molecules may stay for more than a year, while those in the upper layers of the ocean remain for a decade or so. The contents of the Atlantic depths have been there since Bonnie Prince Charlie, while their equivalents in the trenches of the Pacific can be traced back to the days of Columbus. Most Alpine or Himalayan glaciers, solid as they seem, keep their substance for not much more than a century, but those at the base of the Antarctic ice sheet may stay in frozen isolation for two million years.

The wettest place in the world, with twelve metres of rainfall a year, is Mawsynram, in north east India, perched upon the first hills to be hit by the annual monsoon. The driest is the McMurdo Dry Valley in Antarctica, sheltered by mountains, with zero recorded snow. If that is too hard to get to, one can always visit the Atacama Desert in Chile, itself in a rain shadow between two mountain chains, which receives around fifteen millimetres every twelve months.

However extreme such local shortages might be, the overall figures are impressive. Rainfall on land would, in twelve months, fill a cube with each side five kilometres long, the distance from Trafalgar Square to London Zoo, and what pours down on the seas would do that job four times over. Nine-tenths of all evaporation is from the ocean. Some of that blows on to the islands and continents, where it falls as rain. To balance the accounts, the world's rivers discharge twelve times the volume of the Mediterranean each year. Farms, factories and cities now use about a fifth of that, most of which returns, much degraded, to the sea.

The least conspicuous, most transient, and from biology's point of view most important component of the aquatic equation is up in the air. Spread over all the Earth's surface it would make a layer just two and a half centimetres deep. Its molecules stay aloft for not much more than a week before they return as liquid sunshine. Brief as their sojourn might be, it takes just three thousand years for a volume equivalent to that contained in all the oceans to move from land and sea to the air and back. In the global market of moisture, its vapour is the main means of exchange with which the solar output is traded.

Hydrogen and oxygen are lightweight elements, but the union of the two is massive, for a cubic metre of the compound weighs (by definition) one metric ton. Chemicals such as hydrogen sulphide (responsible for the smell of rotten eggs, with two hydrogens attached to sulphur) have a structure rather similar to that of our familiar fluid, but all are gases, rather than liquids, at room temperature.

That physical oddity, water's intimate relationship with sunlight, and its central role in the planet's economy emerge from a detail of its chemistry. The molecule has two hydrogens attached to a single oxygen. The oxygen atoms gain a negative charge and the hydrogens a positive, so that every molecule possesses a north and a south pole. Like bar magnets, they can then stick to their neighbours. One hydrogen in a molecule of H_2O associates itself with an oxygen in another nearby. Each of those hydrogen bonds, as they are called, is involved in about four such relationships, so that in a teaspoon or an ocean the bonds generate a branched network that may stretch for millimetres or miles. If hydrogen bonds were ten per cent stronger, water would appear to be solid, although it would in fact be an amorphous substance rather like glass (which is an intermediate between solid and liquid). Ten per cent weaker, and it would be a gas. Only because their mutual attraction falls within such a narrow range is the compound blessed with so many talents. Each bond lasts for only a fraction of a second, but together they give water much of its power.

First, they keep it in its liquid state at room temperature. They also provide it with a lot of inertia when it warms up or cools down, for the effort it takes to raise the temperature of a piece of iron by 250°C will raise that of an equivalent mass of water by just 100°C and the liquid will take much longer to return to room temperature. This allows water to act as a shock absorber in the Earth's relationship with the sun. Efficiently and without complaint, it redistributes solar energy around the planet in the form of clouds, winds, rivers and currents. Together they damp out what would otherwise be fierce local, annual and daily swings in temperature. The Gulf Stream saves London, for it moves warmth from the tropics across the Atlantic. The British capital lies on the same latitude as does landlocked Calgary in Alberta, where in February only the brave leave their living rooms and where in summer they swelter, but in spite of Londoners' constant complaints, their city has a mild climate.

Within an ocean or a wineglass the molecules are attracted by hydrogen bonds to the same extent in all directions.

Those at the surface, in contrast, have neighbours on just one side, and are pulled inwards. As they are forced to crowd together they generate what is in effect a skin. The effect of this 'surface tension' is to allow small insects such as water striders to skate upon a stream (and I once wasted many weeks in Spain in a study of their sexual behaviour, in which males hold on to females to protect them from other suitors, while their rivals try to kick them off, with more than a thousand attacks needed to dislodge a competitor). Detergents reduce surface tension, force dirt to slide off plates or clothes, and – an experiment furtively indulged in by my students – cause the insects to sink, with dire effects on their sex lives. Surface tension also means that as water falls from a tap or a cloud its drops form spheres.

The bonds that do that job also enable the compound to attach itself to other materials such as glass or metal. That ability allows water to rise within a fine tube by capillary action, a circumstance in which the substance acts more like a gel than a liquid. The effect is universal whenever the fluid touches a suitable solid surface, but becomes most obvious in a narrow pipe. The extent of contact between its internal surface and its contents depends on the radius of the pipe, so that only in the finest channels, with tiny amounts trapped, is capillary action perceptible. The effect might seem no more than a laboratory trick, but is in fact everywhere. It gives us paper towels that soak up spilled tea, not to speak of athletic shorts that suck out sweat. More important, capillary action, combined with the might of the sun, is at the centre of the water cycle, for it enables plants to soak up the stuff from the soil and discharge it to the atmosphere with almost no expenditure of energy. They green, and feed, the world as they do so. The process is called transpiration. Around half of the sunlight that falls on land goes to evaporate water, and much of the job is done by plants as they move it from soil to air.

When it comes to drinking, plants are not like us. They do not recirculate what they imbibe but pump almost all of it into the atmosphere, taking in a small proportion of its dissolved chemicals for their own ends. If a human did the same, he or she would have to pour down forty litres of water – or fifty bottles of wine – each day to stay in balance. Instead our intestines and kidneys soak up most of what we swallow and return it to the bloodstream. In a plant, a small part of its intake is used to run its internal machinery and a little more goes to inflate the cells of leaves and stems to allow them to keep their shape (which is why the leaves of an unhappy pot-plant begin to droop). Most of it, helped by the sun and the wind, pours into the air.

A large oak in a sunny place puts out a hundred and fifty tons of vapour in a summer. It does not get much in return for its efforts, for a ton of the stuff passed from leaf to sky adds no more than a kilogram of new leaves, flowers or timber. Tropical rainforest works even harder, so that every square metre of the Amazon basin emits around eight litres of water a day, much more than does the same area of lake or ocean. To move from liquid to vapour, the hydrogen bonds must be broken. This uses a lot of energy, which means that the surface from which water evaporates, be it leaf or Lycra, cools down, so that wet shirts are less comfortable than dry, dogs pant, humans sweat, and lakes and rivers stay colder than the air above.

Under blue skies, in dry air and in a stiff breeze, plants become fountains of freshness and humidity. A large tree on a hot day can lose enough heat to air-condition two substantial houses. Its trunk contains thousands of tiny tubes that reach to its highest point, which may, in a California redwood, be a hundred metres up. As water evaporates from the leaves, the constant loss draws up the contents of each of the minute vessels below by capillary action, while a smaller amount moves towards the heights by direct passage between the plant's cells. The hydrogen bonds enable their molecules to stick together in a chain, and even in the tallest branches the column has enough internal strength to defy gravity's attempts to break it up. Thanks to sunlight and the laws of physics, it costs the plant very little to carry out this herculean task. Not all the liquid that passes upwards does so thanks to transpiration. Cut the stem of some species close to the soil and it will ooze a bead of juice, as a hint that some kind of pump is at work, perhaps as a safety mechanism to keep itself erect even on cloudy days.

All plants transpire, but they need to ensure that they do not do the job too well. Leaves in tropical jungle are huge, but in dry-land plants such as agaves are reduced to spikes. Other desert species have hairs on the leaves, or a thick layer of wax, while cacti tend to be rounded to reduce their surface area, and use a swollen stem rather than leaves to soak up sunlight.

Most leaves bear thousands of pores called stomata. These are opened and closed with the help of cells on either side that act rather like lift doors. When they are open, vapour gets out and when closed it is locked in. Stomata exert strict control on how much is allowed to escape. In general, they do not open at night, for a plant cannot pump out much vapour without the help of solar energy. Even so, a small amount of the vital fluid is lost even then, perhaps to keep the capillary tubes in working order until dawn appears.

The leaves of some species can track the sun with the help of specialised cells at the base that act as hydraulic motors that bend the stalk. The habit is common in desert annuals, which flower for just a few days after the brief rainy season (if it arrives, that is, for in my half-dozen month-long spring-time trips to Death Valley over a decade or so I saw a 'desert bloom' just once; it was worth the wait). When drought returns, they reverse their behaviour to show the edge, rather than the surface, of their leaves to the skies.

Transpiration from a grass stem, or even a tree, may not seem a difficult task to those who have not tried – like an oak – to carry a ton of water to a height of thirty metres each day, but sometimes the talents of capillary action show themselves in spectacular fashion. Some seeds can force even the most recalcitrant soil to give them a place. Each is filled with fine particles of starch and cellulose. When they get wet, these pick up an electrical charge, which attracts water molecules that stick to them with the familiar bonds. This causes them to swell, and to go on doing so as more and more molecules join in. This process, imbibition, as it is called, can generate a pressure a thousand times greater than that of the atmosphere, and means that an acorn that falls into a crack in a rock can split it apart as it germinates.

The ancient Egyptians exploited that power, for they mined the limestone blocks used in the Pyramids – which may weigh sixty tons – with holes cut in the solid rock into which they hammered thick stakes of dry wood. When water was poured on to them, the bonds worked their magic, the liquid was sucked in, the wood swelled up, and the rock cracked open. The very paper upon which these words are written is held together by hydrogen bonds, which hold its fibres together when most, but not all, of the water used to make it was evaporated during manufacture.

Transpiration is at the heart of the water cycle. The process, together with direct loss from the surface, contributes almost half the vapour that returns to the ground as rain, while in the Amazon basin and its fellows the figure goes up to around three-quarters. Forests speed the cycle in other ways. They release large numbers of particles into the air, so that the clouds above are full of life, with thousands of species of bacteria, matched by a vast array of fungal spores, of pollen and more. These act as nuclei around which – again thanks to those bonds – vapour condenses and becomes a raindrop. In cities the process is hastened by smoke pollution, and in some places the incidence of storms and of hail tends to follow a seven-day cycle as that builds up on weekdays and drops at weekends. Cloud seeding, which uses particles of silver iodide or even common salt, does the same, and is now widely used in spite of some doubts about quite how effective it is. Some years ago, it did seem, at least to the officials involved, to preserve the opening and closing ceremonies of the Beijing Olympics from threatened downpours.

Great streams like the Amazon form their own basins. However grand they may appear their ambitions are modest compared with that of the vapour that rises from the forests that surround them. The sky is full of invisible rivers, whose contents blow for hundreds or thousands of kilometres before they return to earth. A large coastal forest that stretches far inland, such as that of West Africa, acts as a hydraulic pump across which vapour can move for enormous distances. Transpiration, and the cool air and condensation that comes with it, causes atmospheric pressure to drop. That sucks in moist air from the ocean and generates a steady wind that can transport moisture from the coast of a large continent into its heart. Each section extracts its tribute from the ground and moves it into the air, where it is blown onwards until it falls as rain, evaporates, and travels onwards once more. Within the Amazon basin, for example, the trees recycle their own exhalations half a dozen times before the breeze moves them onwards to the Andes, where they fall as rain or snow, fill the continent's rivers, and transform what would otherwise be a desert into a fertile landscape.

The hydrogen bonds show their muscle in other ways. When the temperature goes up, water molecules move apart, so that, like other liquids, when it gets hotter water's density drops. When it cools, it becomes denser again. Then, at 4°C, it has an unexpected change of personality. Below that point, the ability of the moving molecules to disrupt the networks of bonds is less effective than the bonds' own urge to regenerate. Their networks begin to draw in new members. As a result, the density of the fluid decreases as the temperature continues to drop from 4°C towards zero.

At around that point, a regular structure – the crystal known as ice – begins to appear. The molecules are no longer free agents able to swap partners at will, but are forced into an ordered array based on a hexagonal lattice (which is why snowflakes have six sides). To do so they move even further apart, which leads to fewer molecules per unit volume than before. Ice is hence less dense than is its liquid equivalent.

In Calgary, unlike London, in winter the thermometer often drops to minus 20°C. In late autumn, when the lakes that surround the city begin to cool down, the cold surface layer falls to the bottom as soon as it reaches four degrees above zero. When the freeze sets in harder, a thick sheet of ice forms and floats on the surface. It insulates what lies below, much of which stays at four degrees. Those balmy deeps then act as a refuge in which plants and animals can survive the winter. Without the famous bonds, shallow lakes in cold places would turn into slush or ice and their inhabitants would freeze, and in a glacial period every berg would sink and the oceans would turn into solid blocks that might never melt.

When water rearranges its internal architecture to form ice, it expands by about a tenth. The grim terraces of central Edinburgh seem timeless, but some are built of low-grade sandstone. Its pores are large enough for the rain to enter and to force off flakes of their raw material on winter nights

(Aberdeen is greyer, grimmer and even colder, but its granite is solid enough not to let the moisture in). The sole solution is to cut out the damaged section and replace it. To avoid such a fate, statues from Versailles to Moscow are wrapped up in winter. In some places, human ingenuity has taken advantage of such behaviour, for in cold places in China and elsewhere engineers inject water into the coal in open-cast mines, and as this freezes it helps to break up the mass.

If the temperature of pure and still water is lowered very slowly, the hydrogen bond networks extend still further. That impedes the conversion into ice and causes the liquid to grow denser. Such 'supercooled' water does not freeze at zero degrees because its molecules tend to come in groups of four or five, held together by hydrogen bonds, while an ice crystal is based on hexagons. It takes quite a lot of work to move between the two states. A sudden tap on the container to disturb the bonds' equilibrium, or a few grains of dust around which they can bind, will lead to an immediate change of identity. Plenty of reptiles, fish, insects and the like supercool in icy weather, as do some Arctic and mountain plants. The strategy is risky, as contact with a fragment of ice may turn them rock-solid at once. Others cheat, for they have a biological antifreeze. In the laboratory, the Alaskan red flat bark beetle can recover from a temperature of minus 150°C, far lower than any found in nature, while a Canadian frog uses vast quantities of sugar to do the same even when the temperature drops to minus 14°C.

Outside the range of most human experience water inhabits a strange and unexpected universe. A hint of its oddities comes from snowballs. Grab a handful of snow and give it a squeeze. That forces its elegant crystals to cross the boundary between solid and liquid and, as the pressure goes up, the edges of some of them melt. Open your fist, drop the pressure, and that newly released liquid freezes into ice, gluing the missile together, ready to be thrown. Squeeze too hard and, in another change of phase, you get a wet glove and no snowball. In the Antarctic even the strongest grip cannot melt the ice, and snowballs are unknown. Skating, too, has never caught on there, for the frigid climate means that even the high pressure exerted by a fat explorer strapped to a sharp blade cannot push solid ice into the liquid film that every skater glides on.

Under immense pressure and intense cold, the webs of water molecules become so tangled that they generate a substance rather like soft toffee. Cool that down, and even stranger things happen, and more than a dozen crystalline forms of H_2O are now known.

Perhaps the most exotic of all, because it did not exist, emerged in the 1960s and was called 'polywater'. The substance had appeared when clean water was kept for lengthy periods in thin capillary tubes (and, by coincidence, one prominent researcher in the field introduced me as a teenager to my long, and ultimately unsuccessful, obsession with playing classical guitar). Its high density, high boiling point and viscous nature suggested to its discoverers that it was a polymer that might, should it escape, drag more molecules into its grip and turn the oceans into jelly. It was described by the scientific journal *Nature* as 'the world's most dangerous material', but the stuff was, it soon emerged, just ordinary water contaminated with silica that had leaked from the walls of the tube. The Atlantic remained as wet as it ever had been.

A glass of Perrier seems innocuous compared with a vessel filled with nitric acid, but its contents are, thanks to the talents of the hydrogen bonds, able – given time – to etch away almost anything. The liquid can dissolve more compounds than can any other (even if it does less well with oils, fats and plastics, which is bad news for today's oceans, filled with junk as they are).

On its endless journeys, it picks up many passengers, from sand to boulders and from viruses to blood cells. Leonardo da Vinci put it well. For him, water was the *vetturale di natura* – the vehicle of nature. That is good news for creatures made of protein and based on DNA, for it clusters around such molecules and stabilises their structure. It also acts as a lubricant for the transfer of chemical compounds down the chain of reactions that runs life's machinery.

However, given a few millennia, water on its solar-powered journeys becomes the great destroyer. The oceans contain not just ions of sodium and chlorine, but those of magnesium, sulphur, potassium, calcium and bromine, all washed from the rocks. Even gold is present at thirteen-billionths of a gram per litre (which means that twenty thousand tons of the precious metal is out there somewhere). University College London boasts about its chemist William Ramsay's Nobel Prize for work on rare gases, but says less about his failed plan to extract gold from the oceans.

I lived as a child close to the River Leri in Cardiganshire, a stream that seemed to my infant eyes a welcome source of refreshment, but was, unknown to me, polluted by lead dissolved out from abandoned mines upstream, so much so that it can kill cattle who eat hay from fields flooded by its waters. I survived its lethal cargo, but continued to wonder why the river flowed fast, clear and noisy as one climbed up towards their ruins, but slowed to a muddy creek as it reached the sea. And where did the vast peatbog beside its estuary come from? Years later, I walked that river from source to mouth, and did the same for several of its Cambrian equivalents. All had the same lifestyle; an animated youth, a dignified middle age and a slow decline as they snaked across a plain before opening into the sea. They owe their history to the rain, and to its chemical quirks. Every river pays a tribute to Leonardo's claim that 'Water gnaws at mountains and fills valleys. If it could, it would reduce the Earth to a perfect sphere.'

As raindrops become rivulets, streams and rivers, they tear down mountains and build plains. Waves burst into cracks in cliffs and enlarge them into caves that soon collapse. Much the same happens in rivers. Glaciers, too, gnaw at the rocks and break off chunks that are ground down into gravel, sand and mud that move to the plains and the seas. Even a night's hard frost causes the soil to heave upwards, ready to be washed away.

Tall peaks are popular but rare. The average height of the land across the world is eight hundred and forty metres (about that of a modest Welsh mountain). Their summits are renewed as continental drift causes tectonic plates to crash into or override each other. Once again, the hydrogen bonds do their best to help, for the liquid hidden deep within the crust lubricates the movements of the plates, which otherwise would be frozen into immobility.

The Himalayas are, at fifty million years old, the youngest mountains in the world. Their summits are still rising at a centimetre a year, but that will not save them. On their peaks, and everywhere else, in the end the rain always wins. The rivers that carry away their remnants have vast submarine fans at their mouths, the heaped-up corpses of crags long gone. That of the Ganges and Brahmaputra is three thousand kilometres long and one thousand kilometres wide, with a mass equivalent to that of the whole of the modern range. At the present rate the summit of Everest will become a plain in a mere hundred million years. That has already happened in Finland, New South Wales, the South African Veldt and Ontario – very flat, each of them – all of which stand on the eroded foundations of what were once mountains almost as great as the Himalayas. Not by coincidence, each is rich in precious metals extruded from deep in the Earth.

Civilisation needs water, but from the earliest times, the liquid has refused to cooperate. It runs out, or it runs over. Man has long striven to control it, but his successes – spectacular as some might appear – are temporary at best. In the twentieth century, seven million people drowned in floods while ten million died of thirst. The two problems have not gone away, for the amount available per head has decreased by a quarter in the last twenty-five years, while to match that, twenty million people suffer floods each year. The countries most at risk are all in the less developed world.

Lima, the capital of Peru, is almost as big as London, but gets just a centimetre of rain every twelve months. It depends on fitful rivers for much of its supply, while some suburbs have erected nets to condense the fogs that roll in from the sea. Arab countries are the worst off, with one-twentieth of the world's people, but no more than a hundredth of its water. They cope with the help of recycling, imports and desalination plants. Saudi Arabia has the world's largest solar example, which together with dozens of its oil-fired equivalents produces half the country's potable water.

An average American uses around five hundred and fifty litres a day, a Briton around a hundred and fifty, and a citizen of Mozambique about ten. The amount consumed may double within the next forty years, and to meet that demand would need the equivalent of fifty barrages the size of the Nile's Aswan Dam.

Farms are the biggest consumers. Chickens need about four thousand litres for a kilogram of meat, beef cattle four times as much. And how do you fit seventy-five litres of water into a pint glass? It's simple: fill it with beer, for that is the amount needed to grow the barley to make the stuff.

Edward Gibbon, in *The Decline and Fall of the Roman Empire*, analysed what lay behind the disintegration of what had been the world's greatest political union. It began with sedition, and ended with the triumph of the Tiber over the efforts to contain it. He claimed that 'the servitude of rivers is the noblest and most important victory which man has obtained over the licentiousness of nature', for they showed the power of the state. Any escape of the waters from its engineers' control heralded the imminent collapse of a civilisation.

That fate has befallen many other regimes, before and after Romulus Augustulus, Rome's last emperor. The danger of a repeat has not gone away.

The last half-century has been the Age of the Dam and the Ditch. Before 1950, the world had just seven hundred dams taller than fifteen metres. Now it has sixty thousand. Together they have drowned a landscape larger than California and have pushed aside fifty million people. The mountain of concrete has transformed the lives of millions more, but behind the dams and dykes, impressive as they may be, their contents have not abandoned their desire to obey the laws of physics.

Nowhere are their powers more manifest than in the rivers that flow from the Himalayas themselves. In India, Bangladesh and China the passage of water from their heights to the deeps show how the rules of geology change the lives of those who dare to interfere with them and the landscapes upon which they live.

China is blessed with, or tormented by, the world's most rebellious waterway, the Yellow River, in the north of the country. It has long been known both as 'the Mother of China' and 'China's Tribulation', and it deserves both titles, for its riches feed the nation's fields and factories, but its floods have drowned millions. Its contents are laden with fifty times as much sediment as is the 'Big Muddy', the Mississippi. At times it composes a third of the river's volume, so that its flow is in effect liquid mud. In terms of simple output, the Yellow River is not particularly impressive, with less than a hundredth of the flow of the Amazon. Even so, in terms of silt and of slaughter, China's Tribulation leads the world; and, as was the case for the Roman Empire, a failure to control its behaviour has led to the collapse of many dynasties.

Its silt has made its way through several routes from what is now Tibet. Twenty million years ago movements of the continental plates began to push the Himalayas so high that they blocked the northwards advance of the Indian rains. As the storms were held back, drought set in over the high plateau. Its soil dried out and clouds of dust blew eastwards. Over the millennia, they settled on central China and gave rise to a fertile soil known as the loess, the most easily eroded landscape in the world.

As the Yellow River runs through, it gnaws at its banks, washing thousands of tons each year from every square kilometre. When it reaches the plain below, it slows, the mud sinks, and the river bed rises, sometimes by as much as ten centimetres every twelve months. Soon, the force of gravity obliges its course to shift, sometimes with a sudden flood that can move the channel by hundreds of miles in a few weeks, causing the deaths of millions.

Legend tells of a Great Flood some four thousand years ago, and of how the country was saved by the heroic efforts of one man. Part of that story has been confirmed by science. The tale tells of a tribal leader who ordered one of his minions, Gun, to fight a huge inundation. Obedient to tradition, Gun built dams and dykes, but these made matters worse, for once they had filled up they burst. The failed engineer was executed, and his son Yu (born, says the story, from the three-year-old corpse of his father) was instructed to fight the torrent. His approach was to help, rather than hinder, its desire to plane flat the surface of the Earth. He demolished the barricades, and laboured for years to build channels and drainage basins that redirected the Yellow River's flow towards the sea. So impressed were the gods that he was anointed Emperor, and founded the First, or Xia, Dynasty, which began in around 2000 BC.

Much of the First Emperor's work was in the Jishi gorge, quite high in the Yellow River's course. Geology shows that four thousand years ago an earthquake built a natural dam two hundred and fifty metres tall across that narrow canyon. Its remains can still be seen. The river rose behind the barrier for months until it gave way and released fifteen cubic kilometres of its contents, which roared downstream to the coast, two thousand kilometres away, with a flow five hundred times greater than normal. That devastated much of its floodplain, drowned most of its inhabitants and led to a drastic alteration of its course.

The city of Kaifeng now sits on the river's banks, five hundred kilometres from the sea. Its history since its foundation in the fourth century BC is a monument to the might of that stream. The authorities have long tried to control it, but have been obliged to build the dykes higher and higher, so that now the river bed is above the rooftops of many of the city's houses.

Waterborne trade made Kaifeng prosperous, and from around AD 1000 its economy boomed. With a population of almost a million, it was for a time the largest city in the world, but it faced disaster after disaster. Its foundations show how its history has been an endless battle with silt. The modern city stands atop six buried predecessors, with the oldest ten metres underground. Each inundation was followed by political upheaval, an event repeated again and again across China.

A drill deep into its foundations tells of deluges far greater than any in recorded times. Around eight hundred thousand years ago the monsoon changed its path. The rains came in hard, and led to an episode in which the stream poured down at a rate far greater than in recent times. Four thousand years ago came another such disaster, not as dramatic, but still larger than any in the written record. It may well have been the Great Flood that led to the instauration of Emperor Yu.

The people of Kaifeng, and many of their fellows, paid the price for their disregard of the rules of the water cycle. Such carelessness has lasted until today, as hundreds of millions of men and women know only too well. That figure is set to rise, both because of a surge in demand by farms and factories, but more and more because human activity has disrupted the bargain with the sun that keeps, for most of the time, floods and famines at bay.

China has a message for many other places. The nation's extensive archives tell of fifteen hundred floods over the past two millennia, many of them catastrophic. The Revolution of 1949 was itself in part sparked off by the famine that followed the huge inundation of a decade earlier when Chiang Kai-shek ordered his army to break open the defences to 'use water instead of soldiers' to stop the advance of the Japanese invaders. The torrent, freed from its constraints, rushed through until the gap was two kilometres wide. Eight hundred thousand people drowned, four million fled, and vast tracts of farmland became unusable for years. The country has also suffered more than a thousand droughts over the past two millennia, an average of one every two years. In places, the hungry had no choice but to feed on tree bark. In the most severe events, half the population starved to death, and people were forced to eat the corpses of their own children.

Many – perhaps most – of those floods and droughts arose because the ancient treaty between plants and the sun that drives transpiration had been torn up by human activity. The same is true in many other places. The Epic of Gilgamesh of around 2100 BC records an attack on the cedar forest of the Middle East by its hero, king of the state of Uruk, on the bank of the Euphrates, to extirpate a demon. Gilgamesh, in a brief moment of regret, says to his assistant: 'We have reduced the forest to a wasteland . . . How shall we answer our gods?' The gods soon retaliated by turning much of the region into a desert. A millennium and a half later, Plato lamented the degradation of the countryside around Athens: 'In the primitive state of the country, its high hills were covered with soil, and the plains were full of rich earth . . . In the heavens above was a temperate climate and an abundant supply of water. Now, the rich soft soil has all run away, leaving the land nothing but skin and bone.' In the same tradition, Christopher Columbus's son described what his father had seen in Jamaica:

> Every afternoon there was a rain squall that lasted about an hour. The admiral writes that he attributed this to the great forests of that land; he knew from experience that

> formerly this occurred also in the Canary, Madeira and Azores islands, since the removal of the woodlands that once covered those islands they do not have so much mist and rain as before.

The loess in China is among the most damaged soils of all. Once, more than half of it was forested. Then the population shot up until that small section of land housed a quarter of the country's citizens. The hungry peasants cut down the forests and set fires to clear the landscape. They also started to herd goats, notorious for their ability to strip vegetation. By the time of the Revolution just a twentieth was covered by trees. Much of the rest was laid bare, or almost bare, for at least some of the year.

In that decade ten times as much silt as in pre-farming times was washed away. The land became seamed with deep gullies. Peasants were forced to abandon their ruined fields and to move on to pastures new. In time, three-quarters of the loess became a shattered wilderness. The nation as a whole has also suffered, for no more than a thirtieth of its original tree cover remains. Today, China has to feed a quarter of the world's population with less than one part in ten of its land suitable for agriculture.

Across the globe over the past three centuries, ten million square kilometres of natural vegetation have been destroyed, and forest is now lost at the rate of twenty million hectares a year – an area one and a half times bigger than England. A quarter of that is due to the destruction of forest for soy, palm oil or pasture, while the remainder is caused by forestry itself, to changes in local agriculture and to urbanisation. Satellites show that the rate of loss has not decreased in the past fifteen years in spite of many pious words about conservation. Farmland, cities and roads now cover a third of the landscape, and humans control about half of total transpiration. Most of the new farmland is pasture, which releases far less vapour than did the forests and marshes it replaced.

Because a rainforest's green pump works so efficiently, even moderate damage can have large effects on places far away. A loss of no more than a quarter of the Amazon's trees might cut off the river in the sky that irrigates much of inland South America. Its basin contains four hundred billion trees. Even in the late 1960s it had been little damaged, but then the real onslaught began. By the turn of the millennium, half a million square kilometres had been lost, much of it to cattle ranches. The soil has little nitrogen, so that, after a few years, the ranchers move on. The rate of destruction dropped for a time, but the area has lost a fifth of its cover in the past forty years.

There have been severe droughts downwind from it in the past decade, and the length of the dry season in the forest itself has gone up by a week. The twenty million people of São Paulo, four hundred kilometres away, long fed by its vapours, now face a severe shortage. Continued destruction in South America may mean that before long much of its tree cover will disappear. The aerial fountains that pour their contents on to the landscape will dry up and the sun will shine on a vast savannah.

In 2017 several Chinese provinces faced the worst drought

for sixty years. The hinterland of Beijing was hit hard. Large swathes of pastureland were lost, and the deserts speeded up their advance. Even in good times, the city has no more water per head than does Jerusalem. The government responded with a search for new sources, with plans to drill more than a thousand boreholes. The water table around the capital now drops by a metre a year, and in places the ground sinks to match.

That reckless attack on the reserves has been repeated across the globe. Sections of Tehran are sinking by twenty-five centimetres a year while in parts of California's San Joaquin Valley the rate is more than twice that. The farmers have been pumping for irrigation for so long that some places have slumped by eight metres in the past century. This puts roads, pipelines, electricity grids and houses at risk and hints that California's breadbasket may soon run out of water altogether.

Everywhere, the assault has been so furious that a third of the water used by cities, farms and factories is now pumped from below, sometimes with rain that fell ten thousand years ago. California's underground basin loses fifteen cubic kilometres each year, and in India and Pakistan the deposit has been lowered by three hundred and fifty metres in just two decades. A third of the major artesian basins over the globe may fail within less than a century as they sink too far and grow too salty. Such disasters are less obvious to the naked eye than are their equivalents that have dried up and polluted rivers, but may be far harder to put right.

Thales, that unwitting proponent of the talents of

hydrogen bonds, lived at the mouth of the River Meander, in what is now Turkey. In his day the slopes around its course were covered with mature woodlands and the river was filled with reeds (which were used to make the first flute, played by the goddess Athena). The fame of that sinuous stream from earliest times has been eclipsed by its decline. A second-century source complained that 'the Maeander, as it flows through the land of the Phrygians and Carians, which is ploughed up each year, has turned in a short time the sea between Priene and Miletus into solid land' – a process that led to wars over who should own it. Today it faces further deterioration. Near its source, Ovid's 'clearest river in Phrygia' is in places reduced to a rubbish-filled ditch. Further downstream the flow is blocked by dams, and below them the bed lies dry for the summer, testing the claim of Heraclites, who also lived nearby, that no man can step into the same river twice.

The god Meander was depicted by the Romans as a bearded giant with a cornucopia in his arms, but his stream's once fertile banks have starved because they no longer gain their annual bounty of silt, which sinks instead behind the dams. And at the river's mouth, where once stood a famous fishery, pollution has killed off the stocks, while salt water creeps upstream against its weakened flow. Everything may still be water, but from the Meander to the Ganges that liquid is not what it used to be. Sunlight and its servant have built our planet and nourished its inhabitants, but man in his ignorance has pitted himself against both, in a battle that, sooner or later, he is bound to lose.

CHAPTER 4

NO ACCOUNTING FOR TASTE

The universe is nothing without the things that live in it, and everything that lives, eats.

Jean Anthelme Brillat-Savarin, *The Physiology of Taste, or Meditations on Transcendent Gastronomy* (1825)

Carbon is the currency of life. Ecologists and food scientists have long tried to follow its passage from green plants as they use solar energy to take carbon dioxide from the air and pass its constituents to the animal world, whence much of it ends up on the dinner table. That chemical traffic has much in common with man's trade in goods and services, and those who study it often use ideas that come from the world of finance.

Karl Marx was among the first to notice. In *Das Kapital* he laments the plight of children forced to work night shifts in a steel mill: 'in its blind unrestrainable passion, its were-wolf hunger for surplus labour, capital oversteps not only the moral, but even the merely physical maximum bounds to the working day ... It steals the time required for the consumption of fresh air and sunlight'. Charles Darwin, too, saw similarities between human affairs and those of other creatures, and in *On the Origin of Species* wrote that 'all organic beings are striving, it may be said, to seize on each place in the economy of nature'. Marx was an enthusiast for that book, published eight years before his own. For him it presented 'a basis in natural science for the class struggle in history' but, unlike the English naturalist, who opposed the idea that biological progress was inevitable, the German philosopher set out an agenda for a different, and more rational, way of life. Propelled by the scientific rules of society, it was destined to replace all other political systems.

Since Marx's day – and whatever the success or otherwise of the attempts to design a social order to fit his blueprint – there have emerged many parallels between his analyses of the rules of finance and those behind the flow of chemical elements through the world of life. Both are based on chains of exploitation of about the same length. Each needs solar power (either directly or in fossil form), with human demands now so great that a third of the photons soaked up by plants is used to satisfy our appetites, most of it in the form of animal feed. In addition, the economics of each system depends on a complex internal web through which capital, be it gold or carbon, flows, for much of the time, in a flexible and reasonably efficient way. However, both are fatally liable to suffer internal conflicts that can lead to sudden and unexpected catastrophe. For ecology and economics alike, to use Marx's famous phrase, 'merely quantitative differences beyond a certain point pass into qualitative changes'. Biologists and political theorists may not use the same language ('ecosystem collapse', rather than 'revolution'), but through quite different routes they have come up with many of the same ideas.

Marx's own interest in the world of finance began with ecology. He took up the subject in 1842, when a young newspaper editor in the Rhineland. There he learned of the difficulties of the peasants who had long used fallen branches as firewood, but had by a new edict been forbidden to gather them on pain of imprisonment. He accepted that the trees themselves should have legal protection, but it seemed quite unreasonable that dead branches – a common good that would otherwise go to waste or, in biological terms, would enter the carbon cycle – should be denied to the poorest. He complained in an editorial that the policy meant that 'the rights of human beings are giving way to the rights of trees . . . Just as it is not fitting for the rich to lay claim to alms distributed in the street, so also in regard to these *alms of nature*.'

Later in his life, Marx still saw woodlands as a microcosm of the waste and irresponsibility intrinsic to capitalism. He notes that 'the development of civilisation and industry in general has always shown itself so active in the destruction of forests that everything that has been done for their conservation and production is completely insignificant in comparison'. Friedrich Engels, too, deplored the way in which they were treated:

> What cared the Spanish planters in Cuba when they burned down forests on the slopes of the mountains and

> obtained from the ashes sufficient fertiliser for one generation of very highly profitable coffee trees – what cared they that the heavy tropical rainfall afterwards washed away the unprotected upper stratum of the soil, leaving behind only bare rock.

Under Communism, both were sure, such vandalism would not happen. Citizens would realise that they are not the owners of the Earth. Instead: 'They are simply its possessors, its beneficiaries, and have to bequeath it in an improved state to succeeding generations as *boni patres familias*.' However, the state of the modern world still suggests to many of its inhabitants that those good heads of the family have yet to arrive.

Das Kapital is a textbook on the ecology of human production. It describes the manner in which the goods generated by labour travel through a web of exploiters, each of whom creams off some of their substance before they pass them on. That ability to accumulate wealth emerges from the 'surplus value' made available by the workers' efforts and consumed, not by them, but by those higher in the fiscal food chain. The proletariat represented by far the largest segment of society, even if many of its members did not realise as much: 'The bourgeoisie . . . has converted the physician, the lawyer, the priest, the poet, the man of science into its paid wage labourers.' Their efforts, from farm workers to surgeons, were debased as they moved up the pyramid of exploitation. In Marx's day, a pound note given to one of his poverty-stricken Soho neighbours could be a matter of life and death, but for the Duke of Westminster, owner of the Grosvenor Estate on the other side of Regent Street, the sum was beneath his notice. Thousands of men and women toiled to sustain his lifestyle, while neither he nor any of his fellows would ever need to demean themselves by joining them.

The human market in food is a microcosm of how capitalism works. The closer any delicacy is to the top of the food chain, the more expensive it tends to be, both in terms of chemistry and of cash. A hamburger or a battered cod has had a long journey from field or sea to plate. To raise prime beef is costly and inefficient in its own right, but the overheads involved in the sale, processing and preparation of the raw material add another level of exploitation and waste.

Meat is often seen as the prime culprit because of its long supply chain, but fish can be as bad or worse, for those most often eaten are themselves predators. They include not just tuna and swordfish, but also cod and salmon. To consume them is the equivalent of basing the globe's meat dishes on the flesh not of pigs, cows and sheep, but of lions, tigers and wolves. I myself, as it happens, eat fish almost every day. I used to feel smug about that but the habit means that I consume more of the sun's bounty than does someone who prefers chicken.

Food is expensive not just in terms of raw material, but of toil. To provide it takes the labour of one worker in four across the world. The accumulation of surplus value into fewer and fewer hands as the process goes on is impressive. Europe has eleven million farms. They pass on their products to three hundred thousand enterprises in the food and drink industry, which transform them into bread, beer and black puddings. Such items are delivered to three million shops, which sell their goods to half a billion consumers. Both carbon and, as a means of exchange, gold are concentrated as they make their way up the chain. Farm labourers take a small, and diminishing, share of the financial cake, while those at successive levels in the process soak up more and more. People who own supermarkets tend to be better off than those behind the till, and the bigger the chain the richer they get. The Walton family, which holds half the stock in the world's largest, the Walmart group, is worth 150 billion dollars.

Plants are nature's proletariat, for the entire food economy, the Walmart chain included, depends on their ability to mix their labour with compounds in the soil and air and, with the help of solar power, to generate the molecules that flow through every supermarket and every ecosystem. Grass uses the energetic gift directly, grazers at one step removed, predators at two, three, or even four removes. The circle is closed by the humble toilers in the soil or the sewage works, from bacteria to fungi to worms, which break down waste material and allow it to cycle once more. In nature, killer whales and polar bears are at the top of the tree because those creatures eat large fish, or seals, which are themselves predators. On that scale, pigs and people keep close company at around the second level, as they eat both plants and animals. Africans and Asians score relatively low, southern Europe and Japan get a somewhat higher mark, while the burger-loving burghers of North America and Europe are even further up in the pecking order. The piscivores of Scandinavia can afford to look down on them all.

Impressive as markets – in food or anything else – might seem to those who benefit from them, Marx argued that they all rest on 'a whole network of social connections of natural origin, entirely beyond the control of human agents'. Inevitably, he thought, what seems a stable system will in the face of its internal conflicts sooner or later fail (a process hard at work on British high streets for the past decade, with the collapse of one chain store after another).

Marx refers on several occasions to the financial crashes that had taken place in the years before *Das Kapital*. A slump in 1825 had led to the failure of dozens of banks and almost to the downfall of the British economy. It was, in some senses, the first modern monetary crisis, in that it could not be traced to an immediate external cause such as a revolution, a war, or a famine. Various dubious claims lay behind it, among them the existence of a fictitious paradise in Central America which enabled its promoter to borrow large sums against the gold nuggets said to litter its countryside. There have been several iterations of such events since then, the most recent of which, the crash of 2008, still affects society a decade later.

As Marx, who lived when steam was pushing the Industrial Revolution into full flow, also realised, human production, like that of plants, depended on external sources of power, and the amount available determined the society that emerged. *Homo sapiens* was, he said, the only creature that could produce its own means of subsistence. It did so by consuming the products of nature.

First came fire itself, a process confined to Planet Earth, for it alone has the three essential ingredients: fuel in the form of firewood or coal, oxygen, and various natural or artificial ways in which to set it off.

Around a million years ago man began to tame the flames for his own ends. They were used to denature plants and animals by barbecuing their remains, in a process that saves much of the effort involved in digestion and increases the time, and the energy, needed for productive labour. As Marx put it: 'The hunger gratified by cooked meat eaten with a knife and fork is a different hunger from that which bolts down raw meat with the aid of hand, nail and tooth.' The newly empowered economy led to the division of labour: 'The man fights in the wars, goes hunting and fishing ... The woman looks after the house and the preparation of food.' Then came the debasement of one class by another as property was expropriated by its self-proclaimed superiors.

With agriculture and the introduction of draught animals, the power supply leapt up, and inequality increased to match: 'Of all the animals kept by the farmer, the labourer was ... the most oppressed, the worst nourished, the most brutally treated.' All over the world, average height went down, tooth decay ran rampant, and skeletons show signs of starvation and of anaemia. The owners of land and of livestock no doubt benefited, but for most of its participants their lives took a considerable turn for the worse.

With the Bronze Age, some six millennia ago, came another quantum jump. Smelting began and an alloy of copper and tin provided tools and weapons. In time, iron supplanted bronze and, as capitalism became more sophisticated, gold and silver became the means of exchange through which surplus value could be accumulated more efficiently, and more ruthlessly, than before.

There then emerged what Marx described as 'a prime mover that begot its own force by the consumption of coal and water, whose consumption was under man's control'. The first steam engines had arrived. Their ability to replace the work of human hands meant that even more could be soaked up by the exploitative classes as wages were driven down and jobs were lost. Marx illustrates the process by discussing, with rather grudging admiration, the awesome power of the steam hammer and its ability to drive down wages and to put hundreds of workers out of a job.

The James Watt engine, which came into use in 1781, produced about ten kilowatts. By 1800 five hundred or so of the devices had been built, to give the world a generating capacity of about five megawatts. Fifty years later, with the development of more sophisticated versions of his machinery, that figure had risen by three hundred times, and at the end of the nineteenth century some fifty thousand megawatts were being pumped out.

Once again the lower classes paid the price. For the people of industrial Manchester and other cities, standards of living collapsed, as they had at the origin of agriculture. As Engels wrote of conditions there in 1844:

> The race that lives in these ruinous cottages, behind broken windows, mended with oilskin, sprung doors, and rotten door-posts, or in dark, wet cellars, in measureless

> filth and stench, in this atmosphere penned in as if with a purpose, this race must really have reached the lowest stage of humanity.

Half of all new-borns died before they were five and many adults were afflicted with cholera, rickets and tuberculosis. Engels also deplored their meagre diet. The best-paid workers could manage meat every day and bacon and cheese for supper. The less fortunate had meat no more than two or three times a week. Then came the edges of real poverty, in which the only animal food was a small piece of bacon cut up with potatoes, while lower yet there remained just bread, cheese, porridge and potatoes, until on the bottom rung of the ladder, ungarnished tubers were the last resort. When there was no work, members of the proletariat could eat only what was given to them, or what they could beg or steal. If they failed, they starved.

In the modern world, steam engines have been superseded by nuclear reactors. With coal, oil, gas, solar panels and wind farms also at work, output has gone up by more than a thousand times since Marx's day. An average American now needs as much energy as a sperm whale to keep his lifestyle running, while even the British are at porpoise level. *Homo sapiens* has become addicted to electricity, eight-tenths of it still generated by carbon-based fuels.

All this has, as *Das Kapital* prophesied, led to a vast rise in productivity and, for many people, an improvement in living standards (although for most workers in the English-speaking world real income has stalled for a decade, and many, like their predecessors, are forced to survive on junk food and struggle to find a roof over their heads). Marx was right to predict a rise in inequality as more power became available, but he might have been startled to learn that today eight-tenths of global wealth belongs to the most affluent one per cent, and that the eight richest people on the planet together own as much as the three and a half billion who make up the poorer half of the population.

Such inequity – and iniquity – is mirrored in nature. The carbon in all living organisms from bacteria to blue whales weighs five hundred and fifty billion tons (which is, by coincidence, about the same as that poured into the air by smokestacks from the date of *Das Kapital* to the present day). The inequalities – and the imbalances – of the natural market in that substance are as stark as those in its fiscal equivalent.

The distribution of the wealth of the living world as measured in its own currency is highly skewed. Plants hold more than eight-tenths of the total tonnage of carbon, bacteria about an eighth, while everything else makes up little more than a twentieth of that raw material. The whole animal kingdom possesses just one part in two hundred of the element. About half of that is invested in arthropods (insects, spiders, crabs and the like), with another substantial proportion banked in fish. The dose held within the bodies of men and women makes up just one part in ten thousand of life's total investment. However, in terms of the amount of carbon used to fuel and feed their lives, *Homo sapiens* is by far the greediest consumer that has ever existed.

Such dominance comes from the ability to steal solar capital from plants. It began with fire, expanded with farming, and has enabled our own species to become more than a thousand times more abundant than would be expected for a wild mammal of the same size. The 'natural' human population, based on estimates of numbers before agriculture, should be around five million, but is now more than a thousand times greater. With the help of fallen or fossilised firewood just one species has escaped from the ecological constraints that limit all others. Its relatives have paid the price. All other wild mammals when added together now possess no more than one part in twenty-five of the carbon present in man and his domestic animals. Agriculture has transformed our lives, and destroyed most of theirs.

Muesli, beefsteaks and all other foodstuffs owe their existence to a gas that floats free in minute quantities in the atmosphere and is in the end transformed into flesh and blood. Most people (and for that matter some biology textbooks) assume that wheat, daffodils and oaks take their essence from the soil. They are wrong, for as the physicist Richard Feynman once said to illustrate the unexpected nature of science: 'Trees come out of the air!' Ninety-five per cent of the mass of an oak is made from carbon dioxide, a gas which for most of human history made up no more than three hundred parts per million of the atmosphere. Leaves are biological factories that, in effect, burn water to liberate hydrogen ions and oxygen gas and to steal raw material from the skies. Life, like the Duke of Westminster, accumulates the biological capital generated in this way as it flows up the economic food chain. By the time the crucial element has made its way into his ducal corpus (or even my own), its concentration, at one part in six, has gone up by five hundred times compared to that in the air.

Green plants have an intimate relationship with light. Take a piece of turf, unroll it in a darkened room and shine a bright white light upon it through a black and white photographic negative. In a few days, a verdant likeness of a face or a landscape can be seen. Artworks based on the technique tend to dwell on the fragility of existence as the sward dies and the picture fades, but in truth the image is the product of a biochemical pathway that forms and feeds us all. Photosynthesis, as the process is called, uses the green pigment chlorophyll to transform carbon dioxide into useful form.

The first hint of its talents emerged in the eighteenth century. Joseph Priestley, among others, had just discovered both carbon dioxide and oxygen. He noted that the former ('fixed air' as he called it) was produced in large amounts in brewers' vats as they fermented (a practical man, he then invented soda water, a process picked up by a Herr Schweppe, whose descendants grew rich as a result). Priestley also found that mice died when kept in fixed air, but that a plant in that predicament survived and even made oxygen: 'plants, instead of affecting the air in the same manner with animal respiration, reverse the effects of breathing, and tend to keep the atmosphere sweet and wholesome, when it is become noxious, in consequence of animals living and breathing, or dying and putrefying in it.' He did not in fact notice the importance of daylight, but he did express surprise that what was fatal to animals made green plants thrive.

Before James Watt invented his steam engine, fires just cooked food, smelted ores, and warmed people up. Watt's device turned warmth into work. Photosynthesis does the same. It uses the biological equivalents of valves, pressure vessels, turbines and gears to transform one form of energy into another. Without it we would have a tepid but almost lifeless planet.

A first hint of how it does its job came from a clever experiment a century and more ago. A long string of green algal cells was stretched out on a microscope slide in a film of deoxygenated water filled with bacteria. A prism that split white light into its colours was used to illuminate the slide. The bacteria at once moved to two sections of the algal thread, one towards the blue and the other towards the red end of the spectrum. Their thirst for oxygen showed that the gas was generated just by light of those wavelengths. Later there came proof of the existence of two forms of chlorophyll, one for each hue. Neither is much interested in what happens between the red and the blue peaks, so that such light is reflected and the plant looks green.

Both types are needed to harvest the bounty of the skies. The bluish-green form takes the process through all its steps, while the second version picks up a contribution from its own part of the spectrum and transfers it to the first to finish the job. Marine organisms use thirty or so distinct pigments and can as a result use a wider range of the sun's rays than those on land. Each group uses a different part of the spectrum, which gives us the red, brown and green algae, and the blue-green cyanobacteria. In a marine deposit beneath the Sahara are preserved the remains of the oldest chlorophyll of all. They were laid down more than a billion years ago and are a rather startling pink. Whatever their age, chlorophylls share a structure rather like that of other molecules that move gases through the machinery of life. Haemoglobin, for example, has four protein subunits arranged around a central molecule that contains iron. The protein chains change shape as they transport oxygen from lungs to tissues, and carbon dioxide in the opposite direction. Chlorophyll, which is built around a core of magnesium, does much the same.

The compound is somewhat of an arriviste, for the ancestors of today's green plants hijacked it from its single-celled inventors a billion years ago. In a move that would no doubt have given Karl Marx a useful illustration of the wicked ways of capitalism, an ancient and photosynthesis-free cell lineage sucked up a blue-green algal cell and gained at one step the ability to extract surplus value from its efforts to extract raw material for its own ends. The association may have begun as a simple division of labour, but in time one party enslaved the other, which became a skeletal version of its earlier self. What had been an independent organism evolved into a chloroplast, an organelle much reduced and modified to serve the needs of its master. Each still has its own independent piece of DNA as a reminder that its ancestors lived free.

Within every chloroplast is a series of double membranes, arranged in parallel. Embedded within them are groups of around three hundred chlorophyll molecules. The first enzyme in their machinery, ribulose-1,5-bisphosphate carboxylase/oxygenase, known to its friends as Rubisco (biochemists are obsessed by acronyms), is the most abundant protein on Earth.

Feynman, when not explaining botany to his public, was a pioneer of quantum physics, and many of those who study the internal machinery of plants are his intellectual descendants (which means that so are the ecologists who work on food chains, much as that might surprise some members of that rugged breed). Quantum theory has also become embedded in the study of the metabolism of both plants and animals.

The biochemists who teased out the molecular anatomy of the cell long ago saw its machinery on a coarse scale, and over the years identified the nuts and bolts of how it is put together. The study of energy flow from plants to predators is still in that primitive state, but research on the process within their cells has moved far beyond it.

As protons – positively charged hydrogen nuclei – flood in from the sky they are pumped across the chloroplast membrane to set up a charge gradient from inside to out (a mechanism discovered, as it happens, by Peter Mitchell, who was on the staff of my Edinburgh department when I was a student and was later awarded a Nobel Prize). This proton pump is then used to run the machinery that generates sugars, starches and more.

One crucial observation was that in isolated chloroplasts the process can go on at temperatures well below that at which any biochemical reaction could function. That hints that photons – or even a single photon – can, however cold it may be, be trapped and stored for future use. Chlorophyll concentrates such vital sparks until enough have accumulated to do useful work.

As the portraits painted in turf show, photosynthesis has some parallels with film-based photography. On the quantum level it has more in common with the work of a digital camera. That device converts optical signals into their electrical equivalents when a photon strikes a surface and interacts with it to set free an electron. In the camera, the sensor is made of silicon and the electrons stream down a channel, moving to higher energies as they go. They are then manipulated to form the image itself.

As photons enter the chloroplast, they too enter a production line. It moves energy from physics – light – to chemistry, the fabric of every creature. Photons are shuffled up a gradient from one carrier molecule to the next until at last one unfortunate individual is manoeuvred into a dead end. There lurks a specialised form of chlorophyll, present in minute amounts. It draws the particle in with such enthusiasm that the consummation of their relationship is forceful enough to split water molecules and to produce hydrogen ions and oxygen gas.

The first stage of the cycle needs light and releases electrons. These are passed down a conveyor belt that makes the molecule adenosine triphosphate (ATP), which is held as a short-term reserve of fuel in much the same way as a bank holds a reserve of cash. It drives the second phase of the system, which can take place in the dark. That makes sugars, which in terms both real and metaphorical are the bread of heaven (or, given what drives their production, of the heavens). A typical human cell turns over around ten million molecules of ATP every second, which adds up to about our own body weight of the stuff each day.

The main job of the factory in which I once generated modest amounts of surplus value was to produce superheated steam to feed the soap works next door. On the way, the steam passed through turbines that generated the electricity that ran both the factory and the station itself. In much the same fashion, the machinery of photosynthesis involves two complementary systems, the first of which pushes out protons and concentrates them into a pressure vessel, while the second generates electrons (or, as the engineers perceive it, electric current). The heavy machinery of the turbine is no more than a magnified, and less sophisticated, version of the plant chloroplast.

As hydrogen ions build up within that structure, they are forced into minute containers scattered within it. This sets up a gradient in concentration from inside to outside. The ions escape from their prison through small pores. In a mechanical turbine, the pressure builds up inside a sealed vessel, and escapes through a nozzle. Its blades spin and power a driveshaft. Photosynthesis, too, incorporates a set of proteins that spin in response to the stream of particles as it passes. Each full rotation of that biological rotor – which revolves hundreds of times a second – uses fourteen hydrogen ions and makes three molecules of the energy currency.

The main contrast between the engineered and the evolved generators is scale and noise; not the rustle of spring but the scream of machinery (which ruined the upper reaches of my own hearing for life).

Plants make sugars rather than soap. The simplest, glucose, emerges from the green production line. The cell then turns that into useful products. As the biochemists trudged down the pathways that produce organic molecules a series of larger and larger sugars emerged. These are converted into starches, which act as a long-term reserve of fuel. In time, the cell generates amino acids, the raw material of proteins, together with DNA, and of a range of fats, hormones, and more.

Plants are placid creatures, uninterested in brains, muscles and the other talents that make animals what they are. Even so, they trap three times as much of the solar input with their green magic each day as is needed to run the whole of human society. Every second, they generate the raw material of fifteen thousand tons of wood, leaves, flowers and seaweed.

The amount harvested depends on the number of photons available. With the sun more or less overhead, the weather wet and warm for most of the year, and a vast leaf area to soak up its gifts, tropical rainforest is the most productive carbon factory of all, with temperate woodlands no more than half as effective. Rainforest stores about half of all the stocks of carbon held in trees, with up to forty kilograms in a square metre in the Amazon basin. The crops which have replaced large parts of it soak up just a quarter as much, while the pastures that now feed millions of beef cattle are even less efficient. Life in the sea contributes about half of all global photosynthesis thanks to the efforts of a variety of its inhabitants, from tiny cyanobacteria and their relatives, to enormous fronds of kelp, and to corals, which contain green algae that do the job for them.

The energy transformed by plants into usable form in effect defines what life is. Every cell, every plant and every animal has an inside and an outside, with a barrier between them. Their internal world is kept in a more ordered state than the external, but to maintain that privileged position involves a constant struggle against the chaos of the universe they live in. Biology faces a battle against the slide into disorder intrinsic to all physical systems, and to keep it at bay costs a lot.

For more than a billion years after life's origin some four billion years ago the fuel that did the job came from sunlight and from chemistry, as water etched away rock to make methane and hydrogen. More came as deep-sea vents poured hot water into the oceans or volcanoes pumped out lava and gases. In that ancient world biological carbon was produced at a rate a thousand times lower than today. Existence plodded along and got nowhere in particular.

Then, a formative event – the biological equivalent of James Watt's steam engine – boosted the energy budget. The dates are hard to pin down, but the origin of the mitochondrion – invariably referred to as 'the powerhouse of the cell' – came well before that of the chloroplast. An ancestor of all plant and animal cells engulfed a free-living bacterium and used its efforts for its own ends; an event recorded in the fact that every mitochondrion, like every chloroplast, still has its own independent piece of DNA.

Genes were traded back and forth with the nucleus, and over the years the hostage evolved into the mitochondrion, the structure that generates most of its energy. The mechanism is rather like that of photosynthesis, but works in reverse, to burn the fuel that makes life possible. The power generated by the new machinery speeded up their biochemistry, their ecology and their evolution.

Then came photosynthesis itself, and matters became even more energetic. For a time the new mechanism produced just hydrogen, but then the process was picked up and modified by the cyanobacteria, which live on today as blue-green algae. Their efforts gave rise to the Great Oxygen Event that began around two and a half billion years ago. Biology, in effect, caught fire, and – except for the anaerobic creatures killed off in billions by the gas – evolution went into even higher gear.

Bacteria and their relatives the archaea, whose genetic material floats free within the cell, did not join in these manoeuvres. None of them has mitochondria or chloroplasts, and they remain sluggish as a result.

Although they have been around for twice as long as creatures like ourselves and amoebae, both of which are eukaryotes (creatures that hide their DNA within a cell nucleus), bacteria are feeble indeed, with a generating capacity a thousand times lower than that of a typical mammalian cell. They have been slow and unambitious learners as a result. The largest, found in the deep Atlantic, is just visible to the naked eye, while others are far smaller.

Some are essential for our own well-being, while others cause disease, but almost all look to inexpert eyes pretty much the same. Their internal machinery is diverse, and they can live in places such as alkaline lakes denied to more familiar organisms, but on the test of anatomy they fail, for their world has no minuscule equivalents of kangaroos, mice and blue whales, close relatives on the evolutionary tree as laid out in DNA, but quite distinct in appearance. Oak trees, malaria parasites, amoebae and humans also share talents unknown to prokaryotes; sex, development and old age included. Bacterial genes, too, are rudimentary, for most have fewer than ten thousand DNA bases, compared with three billion in ourselves. We eukaryotes owe a lot to the ruthless ancestors who forced their fellows into biochemical subjugation.

All creatures, eukaryotes or not, must, like all economies, compete for their fuel and for their raw materials. Plants need sunlight most of all. To find it they sometimes seem, brainless as they are, almost to make informed decisions. The aspen, for example, flutters its leaves in the slightest breeze. That gives many more of them a chance of a glance at the sun and means that the amount of the crucial gas soaked up increases by a tenth on a windy day.

Aspens and their fellows do not, needless to say, plan ahead in their search for resources, for as Marx (always a useful source of ecological metaphors) put it when he contrasted the work of bees with that of man, an 'architect raises his structure in imagination before he erects it in reality', while bees and aspens do not. Plants, like insects, are lacking in imagination, but they do possess a web of receptors and hormones that provide a simulacrum of that talent. Leaves assess the balance between the amount of red light (useful for photosynthesis) and far-red, not so used. That tells each one whether it is in the shade and, if so, hormones are secreted to make new leaves flatter and thinner and the stem grow faster to lift them above the canopy. Some face an excess of ultraviolet, which damages DNA, and gain the equivalent of a tan, a protective shield that clusters around the cell nucleus while their machinery spins on unhindered.

A plant's raw materials often run short. In a closed commercial greenhouse as the sun goes down, the level of carbon dioxide is less than half that present just after dawn, as so much of it has been soaked up. Their owners sometimes come to an arrangement with local industry to use their flue gases, to their mutual benefit. Other factories do the same for the outside world, for the carbon dioxide that pours from their smokestacks has increased the global growth of plants. Over the past five decades there has been, as a result, a rise in leaf area equivalent to twice the land surface of the United States. Other elements – nitrogen, phosphorus and potassium most of all – may also be in short supply, which is why farmers fertilise their fields.

The chemical mix in soil and air, and the amount of energy available, each help to determine the nature of the community that evolves. In the far north and south, or the depths of the ocean, the levels of sunlight and of nutrients may be low indeed. In the tropics, in contrast, the generous supply of photons and the rapid turnover of carbon render them far more productive than are the temperate zones. That fuels the evolutionary machine. Tropical vertebrates have more diversity than the north and the south. Such places also tend to have more sex and more death, shorter generation times, and a harsher struggle for existence. The wet tropics contain more kinds of plant and animal than does any other part of the world. Half of all mammals live close to the equator. The oceans show the same pattern, with mussels, clams, oysters and their fellows more diverse in tropical seas than closer to the poles. On land and sea, as in the agricultural and industrial revolutions, vital force from the heavens fires up an engine of change.

Land molluscs in their unassuming way hint at its importance. The range of my own favourite stretches from Scotland to the Balkans. The Roman version is common in France and Switzerland. Its relative the brown or garden snail has its ancestral home around the Mediterranean, north and south, while the giant African land snail (which grows to thirty centimetres long) thrives across much of tropical Africa.

Cepaea has been introduced into the Americas, where it is common in gardens and hedges in some places but is more or less harmless. The Roman snail, too, has reached a few distant lands, but has little nuisance value. The brown snail has in contrast been taken from its sunny shores to many places, where it has become a pest. In California it damages the citrus crop and attacks a variety of other plants across the southern states and north to Washington. The animal is also a real irritant in South America, Australia and South Africa.

The giant African species puts all the others into the shade. It has spread across the tropics and causes huge amounts of damage. It eats vast quantities of vegetation and lays several batches of four hundred eggs each year. In Barbados the animal has even become a traffic hazard. Thousands of crushed bodies on the roads turn into a slimy mess that causes cars to skid and kill more, which in turn attracts further individuals anxious to feed on their corpses. As an added delight, because they need calcium for their shells, the creatures often get into houses and eat plaster from the walls. The contrasts among the four species find their roots in the metabolic strategies of creatures that evolved to cope with quite different levels of solar input, from cloudy Scottish sand dunes to sunny Mediterranean shores and torrid African jungles.

Every one of those creatures depends for its raw material on the ability of plants to turn a simple gas into complex chemicals each of which in the end returns to the air and is recycled, on a timescale that varies from minutes to millennia.

The carbon factory in the Amazon jungle operates, like a modern car assembly plant, on a 'just in time' model, for the stock picked up and processed has no more than a brief stay within its host before it returns to the air. Other elements also pass through at such speed that the soil has few reserves, and most is held in a short-term account, the trees themselves. In the hot and steamy atmosphere, with the earth full of insects, bacteria and fungi, any wood that falls to the ground decays fast, and its contents return to the air or to the food chain.

That strategy is efficient but risky, which may be why such places are so fragile. When cut down for crops or for pasture, the new fields – like the coffee plantations in Cuba criticised by Engels – may last no more than one or two seasons before the soil is exhausted and the farmers move on.

A larger deposit of carbon resides within the sea of conifers that stretches across Canada, Alaska, Mongolia, Kazakhstan, northern Russia and Scandinavia. In spite of its sunless winters and cool summers, when all its reserves are included this green assemblage holds a stock twice as big as that of the tropics. Its work – unlike that of the flashy world of the equator – is slow but steady, for its trees deposit enormous quantities of the mineral in the soil, where it can stay for years, while the peat bogs and marshes that surround them do the same. Given time, peat itself can be transformed into coal.

Once a plant has been eaten by an animal its raw material enters the wider biological economy. Its members – like those in the human food chain – accumulate it as its molecules move upwards, from grazers to top predators. Each time a packet passes from one consumer to the next, the amount available for the next in line drops, because some of its worth has been extracted by the previous holder to maintain itself and to reproduce or, if it dies, to return its contents to the base of the chain. In most modern financial systems, a shopkeeper, a trader or investor gains a few per cent each time a deal is made. A participant in its natural equivalent on land passes, on average, no more than one part in ten of the minerals it receives to the next link in the chain.

Ecologists know less about the economics of the natural food chain than do investors in supermarkets. The old story was that 'big fleas have little fleas upon their backs to bite 'em, and little fleas have lesser fleas, and so on, *ad infinitum*'; grass soaks up sunlight, deer eat grass, and wolves eat deer. Surplus value moves upwards in a simple linear fashion.

Snails show that such a model is far too simple. The giant African version is a pest in many places, while the Central American Rosy Wolf snail is a voracious cannibal that eats other snails and slugs. The creature, almost as big as its intended victim, was introduced into many Pacific islands in the hope that it would kill off the African invader. It seemed obvious that the cannibals would attack the most abundant and substantial item available but they had not read the textbooks and turned most of their efforts towards the unique and much smaller indigenous molluscs, driving many to extinction. On a visit to Hawaii a few years ago a colleague and I spent days in a search for the remnants of a group of native snails which had once been abundant. We found just one individual. The cannibals have destroyed what had been stable ecosystems for thousands of years. They will never recover.

Food chains in their complexity are often studied in seas, lakes and rivers, both because they are easier to sample than those on land, and because the fishing industry is very interested in the results.

The first hint of the power of accumulation in the watery world came not from a food, but from a poison. The carbon-based nerve agent DDT was introduced in the 1940s. Large amounts were used by American farmers, but dire effects soon emerged.

The substance was washed from fields into lakes. There, it had a concentration of one three-hundredth of a part per billion. It was picked up by single-celled creatures and plankton, which are eaten by animals such as water-fleas. These are then devoured by small fish, which themselves fall prey to larger fish such as trout. Predatory birds are at the top of the chain. At each step the compound is concentrated. At the top, ospreys contain a concentration eight hundred times greater than in the water – a level so high that their eggs failed to hatch and populations collapsed. Rachel Carson's *Silent Spring* of 1962 led to public concern, and ten years later the molecule was banned. For the carbon held in sugars or proteins as they move from green algae and their fellows to tuna or albatrosses, the figures are much the same.

Marine consumers vary in size by a factor of twenty or more, from tiny plankton to blue whales. The food chain from bottom to top depends, like all others, on photosynthesis. In the sea, the top metre of water absorbs more than half the sunlight that gets in. No more than a fifth of it reaches down to ten metres, with all of it taking place within a hundred metres (and sometimes much less) from the surface. The biggest players are the phytoplankton – tiny creatures equipped with green and blue-green pigments that soak up sunlight and turn over carbon at ten thousand times the rate of trees on land. They in turn are eaten by tiny animals – the plankton – many of them in the form of new-born fish, crabs and the like. These face swarms of larger crustaceans, some of them, the krill, present in enormous numbers in the Antarctic, where they are the main food of both blue whales and tiny fish. They also fall victim to larger attackers such as anchovies, which are then eaten by smaller predators such as mackerel, which become the victims of their grander kin, tuna and swordfish included. Seabirds (which catch as many fish as do the world's fishermen) add another link to the chain.

Some of its participants were scarcely known until a few years ago. Sonar was invented at the time of the First World War as a tool to detect submarines, and later in attempts to map the ocean floor. The first versions had mixed success, for their signals could not penetrate further than about a kilometre below the surface, and for unknown reasons did much worse at night than in daylight. In fact, the sound waves were blocked by the creatures of the 'twilight zone' – the marine layer two hundred to a thousand metres down into which some light penetrates, but too little to allow photosynthesis. In daylight hours its inhabitants stay in the deeps, but at night they move upwards to feed.

The most abundant among them are small fish from two to thirty centimetres long, members of a group called bristlemouths. They have luminous green or red spots but are otherwise black. They are the most abundant vertebrates on Earth, with quadrillions, thousands of trillions, of animals. They represent nine-tenths of the weight of all fish and a yield a hundred times that of man's annual harvest from the seas.

Nobody has yet worked out a way to catch them, but with

more than a ton for every man and woman on Earth they could provide a useful source of protein. They already do a valuable job on our behalf, for they are enthusiastic defecators that produce a torrent of carbon-rich waste that falls to the bottom, where it will stay for centuries. The bristlemouths and their relatives eat small crustaceans and are in turn eaten by squid, cuttlefish, and fish such as anchovies and mackerel. The denizens of the twilight zone are hence key players in the marine food chain.

A closer look at the efficiency of carbon transfer in that system suggests that the 'one part in ten' rule – the claim that just a tenth of that currency is passed on at each step in a terrestrial food chain – is too simple. In the sea, when a portion of carbon passes from phytoplankton to small crustaceans one-eighth gets to the predator. When a shark eats a lesser fish the process is more wasteful, for just half of that proportion makes it. Those figures show why those that reach the top of the tree become such an expensive luxury. A tuna needs ten thousand times its body weight in terms of the tiny creatures at the base of its food supply before a slice of its flesh appears on a plate, while a warm-blooded predator such as a killer whale or a gannet is even pricier.

All this is reflected in economic terms. Rich Japanese have paid a million pounds for a single prize tuna. In China the shark fins popular among the affluent use no more than a tiny proportion of each animal's goodness, for the rest of the corpse is dumped into the sea. The fins sell for around six hundred and fifty American dollars a kilogram, which is about three times the record amount paid for an equal weight of tuna. Such figures show how human tastes may sometimes verge on the irrational.

The passage of carbon through ecosystems on both land and sea is better understood than it was, but what determines the shape and the length of each section of its tangled web? One of the few general rules is that food chains are on average about one step longer in the sea than on land (which makes a taste for shark fins even more wasteful). But why? There is no consensus. Almost everyone seems to have their own explanation for their own favourite system. Chains are longer in warm places, or they depend on the presence or otherwise of predators. They are bottom-up, with the amount of photosynthesis the crucial factor, or they are top-down, with external disturbances such as storms or fires in charge. Other models concentrate on 'pinch-points', the notion that changes in the number of individuals in one of the levels – in the sea perhaps the bristlemouths – control the whole system. One widespread idea has it that the amount lost at each step as one ascends the chain is the most important factor, while yet another view is that the size of the ecosystem counts, so that small lakes or islands have shorter chains than do large.

All these processes may be at work, perhaps several of them at a time, and it may be that their relative importance varies with habitat, or with the presence or absence of predators. Even so, the science of carbon transfer through communities resorts too often to special pleading. In that it resembles a once popular, but now discredited, approach to the study of the passage of the same element through cells.

As mathematicians sometimes say, 'If I have seen further, it is by standing on the shoulders of Hungarians,' and even if in mathematical terms I have not seen far, an Edinburgh-based Romanian of Hungarian descent gave me a leg up. Henrik Kacser – short, bald and always wreathed in smoke – gained our attention in his first lecture when he pinned up an image of the cell's metabolic pathways, the road map of a set of chemical cycles which students of biology then learned like novices reciting the catechism. He tore it into pieces and announced that: 'None of this is worth knowing. All you need is to ask what goes in at one end and what comes out at the other.'

A few years later he published a mathematical model that founded what has now become the considerable science of systems biology. It points out that: 'The properties of a system are in fact more than (or different from) the sum of the properties of its components.' Traffic through a cellular pathway was once assumed to be controlled by lights set at red, green or amber. Biochemists searched for 'rate-limiting steps', supposed bottlenecks in the journey. Large amounts of money were spent by brewers, bakers, drug developers and agricultural researchers in a search for choke points that might be targeted to increase productivity, improve yield, or cure disease. None were found, because they do not exist. A cell is not like a factory, a series of separate production lines, but a complex web through which raw material passes in a multitude of ways. A soap works may be efficient, but it is both linear and fragile, for the failure of just one turbine may close down the whole operation. Life, from cells to rainforests, is in contrast profligate, flexible, and remarkably tough.

Kacser suggested that biochemical compounds flow through cells much as cars travel from place to place. On a motorway they are unhindered by traffic lights and they overtake and are overtaken, with constant changes of lane so that the system is much more effective than would be the same space used for three separate one-track roads. An accident that blocks the carriageway can be circumvented with a diversion down a slip road. What matters is to arrive as soon as possible, rather than the details of how to get there.

The internet is Kacser's notion made electronic flesh. The system is all, and the stream of codes that makes up a brief email may take a dozen tracks across the globe before it comes out at the other end. Nobody cares whether different bits of the message have gone via Virginia or Vladivostok, and indeed there is no way to find out quite how they travelled. Today it provides a simulacrum of culture at the touch of a keyboard, but it was invented to give flexibility to military communication, so that even if great sections were wiped out by nuclear explosions, it would still function. In 1973, my own institution, UCL, was the first outside the United States to have a link to ARPANET (the Advanced Research Project Agency Network), and that fragile transatlantic bond has now expanded into a global labyrinth.

In its very early days, not much more than four decades ago, local network collapses were frequent because of a shortage of pathways through which the data could pass, and even now they are not unknown. The number of routes is predicted to double in the next couple of years, but even that may at times not always be enough to keep up.

The systems theory upon which the World Wide Web, the electricity distribution network and the train timetable rest is now much used by brain scientists, geneticists and students of cell division. It has also become important to ecology. More and more, those who study food chains are beginning to ask if there are any firm rules at all behind the flow of carbon. In fact, such networks are, as Kacser would be happy to learn, not simple production lines, but complex and interlinked systems with many fail-safe mechanisms built in.

A closer look shows some unexpected parallels between their machinery and that of cells. The numbers of individuals of a particular species in a reef or a beech-wood and the numbers of molecules of different enzymes in a cell have a common pattern – a few are very abundant, quite a lot rather frequent, and a huge number are almost never seen. The links among them also look much the same. In yeast a path with just three or four junctions links most of the cell's products – alcohol with the proteins of the cell wall, for example. In the forest or reef, too, most food webs have just four or five steps, be it from mushroom to eagle, or from seaweed to shark.

Another hint that cells and forests have a lot in common comes from the discovery – that as in ARPANET – great sections of each can be destroyed to rather little effect. Geneticists can 'knock out' genes in yeast or mice in the hope that changes in their appearance, their health or their behaviour may hint at what the missing section of DNA actually does. That has been a useful tool, but the surprise is that for many of the experiments, a mouse or fruit-fly seems not to notice even when what seems a major link in the biochemical chain has been removed. Plants are even more resilient, for no more than a small proportion of individuals in which different genes had been extirpated show any change in their well-being. In yeast, every gene they possess has now been knocked out to see what happens – and for about half the answer seems to be, as far as we can tell, 'nothing'.

Much of that robustness rests on the existence of alternative pathways through which molecules can flow, together with spare copies that come into use when things go wrong. Cells too are more resilient than once imagined. Cancer shows how such resilience can collapse, for human DNA can take hit after hit with no apparent response – until suddenly the cells involved turn into aggressive tumours as the internal defence network can no longer cope.

Much the same is true of economies, of forests, or of coral reefs. They can persist through bad times and good, but for all of them there is a limit: years of abuse seem to do little harm until, one day, everything dies.

Many creatures, from lynx to lemmings, show cycles of abundance, sometimes by as much as a hundred times. For a few – such as the joint fluctuations of lynx and their snowshoe-hare prey – we know what drives them, but often, they have a life of their own. Grow flour beetles for many generations on a standard dose of food, and the numbers stay about the same, but suddenly, for no apparent reason, they go in for wild swings in abundance that last for generations and may end in extinction. The same is true for small lakes in Holland, for some have clear water and green plants, while others are murky. Now and again a particular lake will switch from one state to the other with no notice, only to change back after a few years. Plant pests such as caterpillars may also become a plague for one or two summers, but then disappear for decades. The natural world also faces repeated and unpredictable physical disasters – fires, floods, storms, and more – which may appear to wipe out a whole ecosystem but, almost miraculously, may be resurrected. Sometimes, though, it is not, so that many patterns in ecology fit Marx's description of the inevitable fate of society: 'All that seems solid melts into air'.

At first sight, nature has a remarkable capacity to recover from such skittish behaviour, whether it results from internal upsets or external threats. Forests are often destroyed by fire or tempest, but they have an impressive ability to rise from their own ashes. Even in the most devastated terrain, enough remains of the original inhabitants to allow the landscape to reassemble itself. What emerges may differ in its details from what went before, but is still recognisably the same habitat.

Robust as it might appear, a forest lost to flames or to storms needs to retain enough of an identity to enable it to be born again. A small, or even a medium-sized patch may just by chance lose a crucial link in its food network, and will be unable to regenerate because it has no fall-back pathway. That fragment will then disappear. In a larger tract, in contrast, the system is big enough to ensure that it is more than the sum of its parts, as damage in one section can be compensated for with an import from a neighbour.

An inadvertent experiment in a mountainous Thai jungle showed the process at work. In 1987, a large valley was flooded to make a reservoir. As its waters rose, they created dozens of new islands. The rich habitat around its shores continued to thrive, but on the islands, in their isolation from the larger ecosystem, matters were less hopeful. Within five years the smallest had lost most of the mammals that once lived on them. Larger patches did better, but within a quarter of a century almost all had gone. On most of the islands there now remains just one survivor, the invasive Malayan field rat.

The same process has happened in many places and on many scales. The Atlantic Forest of Brazil lies south of the Amazon and stretches to the coast. Half its plants and the same proportion of its mammals are found nowhere else. It was, in the nineteenth century, still more or less in its primitive state, its sunny glades full of life. In February 1832, Charles Darwin arrived there on his landfall in South America:

> The day has passed delightfully. Delight itself, however is a weak term to express the feelings of a naturalist who, for the first time, has wandered by himself in a Brazilian forest . . . To a person fond of natural history, such a day as this brings with it a deeper pleasure than he can ever hope to experience again.

Were Darwin to return, he might be disappointed. When the Portuguese invaded in the sixteenth century the forest covered more than a million square kilometres and stretched far inland. Nine-tenths of it has now gone, but an area almost the size of England is still left. That might seem enough to allow it to persist, but the depredations of farmers and loggers have divided it into hundreds of fragments. Most still retain quite a lot of the diversity – tamarins, spider monkeys, sloths and more – that so impressed Charles Darwin. Even so, the insidious truth is that their plants and animals have begun to fade away at a rate that depends on the area of each patch. In time, like the islands in the Thai reservoir, the whole ecosystem will fall apart.

The ocean shows the process at work on a much shorter period and an even larger scale. Steller's sea cow was known to science for just a brief interval before it was driven to extinction. The creature was first seen in 1741 by the German naturalist Georg Wilhelm Steller when he was shipwrecked on an Aleutian island on the way back from the first Russian expedition to Alaska. It resembled a modern dugong, but was twenty times larger. The animal ranged in vast numbers through the kelp beds of the North Pacific, eating their fronds. Unfortunately for the sea cow, a couple of years later, thanks to information brought back by Steller, the Russians began to hunt sea otters for their fur along the Pacific coast of the Americas from Alaska as far south as California. Within a decade and a half the otters were almost extinct. Otters eat sea urchins, which boomed as a result. The urchins graze on kelp and consumed it with enthusiasm. The beds of giant seaweed died off and a long stretch of once productive coast became a rocky wasteland. The kelp had been the food of the sea cow, and the last member of that species died on a remote

Arctic island just twenty-seven years after its discovery. The creature is the icon of extinction and the paradigm of how to ruin an ecosystem in one step.

Two and a half centuries later, the ocean's ecological economy has once again collided with its fiscal equivalent, with severe damage to both parties. Top predators – tuna, cod, sperm whales, and more – have lost nine-tenths of the numbers that were present before industrial fishing began, a century or so ago. Half the decline has taken place in the last fifty years. Giant vessels pillage the seas. They hoover up vast numbers of fish, and to make matters worse, one in three does not make it to the table, for it is thrown back as unprofitable or beyond an official quota, or because the catch spoils before it can reach the market. The efficiency and the cost of the trade each continue to increase, but overall production has not gone up for thirty years, and we are close to, or perhaps even beyond, 'peak fish', the level at which the reserves have been depleted to such a degree that the whole marine food web may be on the edge of failure.

When the Grand Banks fishery off the coast of Newfoundland was discovered by the Portuguese in the sixteenth century it was claimed that a basket lowered into the water would always come up filled with large cod. A century later the English colonists of New England noted that the fish were still so abundant that 'it seemed that one might goe over their backs dri-shod'. In the 1960s its waters were invaded by trawlers that could scoop up as many fish in half an hour as a Portuguese ship could have managed in a season. Peak cod came in 1968 with a harvest of almost a million tons, but within less than three decades the yield had collapsed and the fishery was closed by Canadian government decree, with a loss of tens of thousands of jobs. It may not be ready for revival for another decade.

The world has seen four hundred such catastrophes in the past half-century and at the present rate there will be no exploitable fish at all left in the Asia-Pacific region within thirty years. Now the industrial ships have been forced to hunt not just for tuna or cod, but for anchovy. Even then they still struggle to make a profit, and some have turned to krill, the food of the great whales, creatures just one rung up from the plankton. Krill are now eaten by many Japanese, Koreans and Russians (and the famous Noma restaurant in Copenhagen, perhaps in a subliminal warning about the future, even offers a plankton cake as a final course). When it comes to seafood, men and women have been obliged to forage further and further down the food chain. In the past twenty years, tuna too have shifted their diet from mackerel and other large fish in the sunlit upper waters and now have to dive to the twilight zone and its innumerable bristlemouths. They have been forced by human greed to join the race to the bottom.

In another turn of the ratchet of ecological irresponsibility, half the world's catch is turned into fishmeal to feed domesticated predators such as salmon, whose ancestors caught their food for themselves. Each year, millions of tons of mackerel, sardines and anchovies go to feed fish rather than people. Salmon production began in Scotland in the 1970s. The yield in the first year was fifteen tons; it is now a thousand times more. Even tuna are now held in large nets in the ocean and fed with squid, krill and the like to let them grow to a decent size before they are killed. Many of those sold in Japan come from such cages, moored off the coast of Australia, or in the Mediterranean. The animals are captured while immature and never lay their eggs, which depletes the stocks still further.

It is not compulsory – and at this rate of expansion it may soon become impossible – to eat tuna or salmon in order to survive. In some places, humbler marine creatures once sustained a whole way of life and might do so again.

Once again, molluscs help tell the tale. In California I worked for a while on one of the distant Channel Islands, San Miguel, on a fruit-fly release experiment that was such a failure that I have quashed most memories of it. One that abides is of the mountains of shells of abalone, mussels, limpets and the like left on the chain of islands by the Chumash, the Native Americans who lived there for ten millennia before they were removed from the islands in the early nineteenth century. In the mainland missions to which they were transported they were soon driven to extinction (and a couple of centuries later fishermen did much the same to the abalone). In the Chumash homeland, the marine snails provided almost ninety per cent of their protein, with fish, seabirds and dolphins making up the rest. Man, that shows, can live almost by molluscs alone and in many places, and for many years, he has done just that.

It was, said Jonathan Swift, a bold man who first ate an oyster. He was himself fond of them, and not just for their taste:

Your stomach they settle,
And rouse up your mettle:
They'll make you a dad,
Of a lass or a lad;
And madam your wife,
They'll please to the life;
Be she barren, be she old,
Be she slut, or be she scold,
Eat my oysters, and lie near her,
She'll be fruitful, never fear her.

That doggerel has, unexpectedly, gained support from science, which hints that a taste for sea snails once saved mankind from starvation, or even extinction.

The bold man who first essayed such creatures lived not in historic times, but around a hundred and fifty thousand years ago, in the Middle Stone Age. In Africa some fifty thousand years before that, there had set in a long cold and dry period which turned much of the savannah and grassland upon which the immediate ancestors of *Homo sapiens* had hunted into sterile deserts. Famine killed vast numbers of people.

The evidence lies in the genes. Their descendants, ourselves, have a rare qualification, for in genetical terms we are the most tedious of all primates, with no more than a fraction of the inherited diversity found in chimpanzees, gorillas and the rest. The reduction is a relic of a major crash in human numbers, perhaps to just a few hundred, at just that time in history. Most of inland Africa was abandoned, and the small band of survivors moved to the coasts of what is now South Africa, one of the most productive shorelines in the world.

There, they found a new source of food. Many of the caves in which they sheltered from the cold, together with the middens into which they threw waste, contain large numbers of shells of mussels, limpets, abalone and their kin, while the chemistry of their inhabitants' bones shows that their diet had moved from the determined carnivory of their predecessors to one based on the sea, on shellfish most all.

From the moment when he took up this new way of life – and in an era in which he was still an endangered primate – man began to become modern. His brain grew fast, he made more complex tools, and quite soon started to experiment with technology and with art. The arrow points found in the caves are made of a stone called silcrete. It has been heat-treated to increase its hardness, a hundred thousand years before that method was re-invented in France. Beads made from seashells made their appearance, and ochre, a red pigment, has been found stored in abalone shells in those South African refuges. The oldest known artwork of all – a grid of ochre lines painted on to a fragment of rock seventy-three thousand years before the present – also comes from one of those caves.

We are the only primate to eat shellfish, which were the last addition to the human diet before the origin of farming. Unlike their hunter-gatherer predecessors, and their agricultural successors, the cave-dwellers' skeletons were well supplied with important trace metals such as iron, iodine, copper and zinc, which were often deficient in the heart of Africa. The mussels, limpets and abalones upon which they feasted are also filled with omega-3 and omega-6 fatty acids, which cannot be made in the body and are essential for brain development and growth. A pregnant woman can obtain enough of those crucial compounds by eating just one mussel a day (which would perhaps not have surprised Swift). Not only did the marine molluscs save our endangered ancestors from starvation at a critical moment in their history, but they may have done a lot to make us what we are, most of which depends on the benefit of the biggest brain in relation to body size of any of our relatives.

Humans took their first tentative steps out of Africa around a hundred thousand years ago. Many of the migrants did not get far, but in time more and more emerged, to fill much of the Old World, and then the New. Much of that era was an ice age, with sea levels far below today's, and much of the landscape impenetrable. The travellers had little choice but to follow the coast – and no doubt they too were fuelled by oysters, abalone and their kin on the way. We owe more to those creatures than perhaps we realise.

In Britain, with its long coastline, the oyster remained a popular dish until not long ago; as Mr Pickwick's servant, Sam Weller, says: 'It's a wery remarkable circumstance . . . that poverty and oysters always seem to go together.' In the mid-nineteenth century half a billion oysters passed through Billingsgate Market in London. Thirty years later the number had collapsed. Even so, an eminent biologist – Thomas Henry Huxley, Darwin's Bulldog as he was known – recommended that given the enormous numbers of eggs produced by the animals, no control on exploitation was needed. That was a mistake, for the numbers soon fell to almost nothing. As a result, oysters are in Britain now a symbol of wealth rather than poverty.

Across the Channel, things were different. The shore was government property and was rented out under strict conditions, with natural beds open sometimes for just a few hours a year. They were supplemented by artificial hatcheries that spread around the coasts. France, as a result, now produces two hundred times as many oysters as does Britain and, as the animals need no artificial food, the business is both profitable and ecologically sound.

When it comes to the oyster's cousins on land, the contrast between the two shores of the English Channel is even starker, for this country continues to avert its gaze even as its neighbour delights in them. 'Wallfish', to use the dialect word, were once a delicacy in Somerset, but neither the historical records nor archaeology give any sign that snails were ever important elsewhere in these islands. In many other parts of the world, in contrast, they have been, and still are, a staple part of the diet.

The French now eat half a billion snails each year, well ahead of Italy, Spain and Germany. All those countries have festivals that celebrate them, and in France at least a certain amount of ceremony still surrounds a meal of *escargots*.

Britain is free of such jollity, and the local molluscs have long been left unmolested. That is odd, for these islands have plenty of places – sand dunes, sea cliffs, chalk downs, nettle-beds – in which they are abundant. Quite why our diet was – and is – so limited nobody knows, but it is a sign of a gastronomic conservatism that began long ago. That attitude is unfortunate, because it would make a lot of financial and ecological sense to eat them rather than to insist on beef or tuna. A hunter uses a great deal of energy to chase down, butcher and roast a deer or a mammoth, as, indirectly, must someone who buys a ham sandwich. A plate of snails can in contrast be gathered and cooked with almost no effort. They are also far more efficient in their use of plant resources than are sheep or cattle.

The real Stakhanovite is the giant African species. Each animal contains more protein than an egg, and is low in sodium and cholesterol. Many are now grown in simple corrals made of old truck tyres and fed with scraps of food and vegetation, while large farms run on scientific principles have begun to spread. So many are produced in Thailand that the mountain of shells has become a real problem. In China, too, they are a popular food.

In Britain, snails are still reviled but, in secret, and at long last, they have begun to slither on to the menu. Because overfishing has reduced stocks, the price of fishmeal has risen fourfold since the millennium. The giant African mollusc is in some places now used to make a powder that is used to feed farmed fish, chickens and prawns. As a result, many Britons, disgusted as they might be by the idea of eating the animal itself, now do so at second hand as they enjoy their fried chicken, smoked salmon or prawn cocktail.

As they consume the remains of a creature just one step from the plants that generate the raw material of the carbon market, the British have moved their position in that system downwards by a tiny fraction. If our fellow citizens could be persuaded to cut out the intermediates in that food chain and cook the snails themselves, the collapse of nature's economy might, perhaps, be postponed for a few more days.

A mass conversion to a molluscan diet in these islands, even one based on oysters and mussels, seems unlikely, but other changes are now beginning to make a real difference to the shape of the food economy. About a fifth of all global deaths each year can be ascribed to poor diet, a proportion that outweighs those due to smoking or to high blood pressure. The best predictor of early mortality is salt, a major ingredient in processed meat dishes. A shortage of whole grains, fruit and vegetables comes next in importance; and the effect is seen even in developing countries, where obesity is becoming an epidemic.

Across the world, people have started to notice, and to move down the food pyramid. In the last fifty years, its summit has become increasingly crowded, for the amount of meat eaten in North America has doubled, in Africa it has risen by four times, and in Asia by a factor of fifteen. Despite that growth, we may be close to, or even beyond, peak meat. India has half a billion vegetarians, and if one believes what people say when asked questions about their diet, Britain too can count more than six million with the same habit. Even more are 'flexitarians'; they eat an almost meat-free diet, with occasional lapses. In the developed world, a switch to an entirely plant-based diet, unlikely as that is, would have a dramatic effect as it would reduce the need for arable land to just a quarter of its present extent.

A hectare of land devoted to edible plants would produce twenty times as much protein as would the same area used to rear beef cattle. Pork is almost as wasteful, while chicken would generate about half the protein obtained from an equivalent weight of field crops.

Vegetarianism is spreading fast. Until not long ago, China saw pork, once a luxury, as a staple. For reasons of health, to reduce the resources needed, and to use land more efficiently, the authorities have now set out to cut its consumption by half. The figures are already dropping, and there has also been a move from that flesh to fowl. In the United States, that land of carnivores, peak beef came forty years ago, and consumption has declined by a third since then (some fast-food chains now even produce soya-burgers that ooze beetroot juice that looks like blood). Americans still eat plenty of flesh, but nowadays almost half is chicken. So great has been the move to that staple across the world that those birds now represent seven-tenths of all avian life on the planet.

The retreat from meat is, for reasons both ecological and economic, bound to continue. The cost of production in terms of the destruction of forests for feed crops, the pollution of land and water, the need for fertiliser, the generation of methane and the coming rise in the human population all make it clear that our present diets cannot long be sustained. To prevent starvation in the face of population growth, across the world the average citizen will, by 2050, need to eat no more than a quarter as much beef, a tenth of the amount of pork, and half the number of eggs as today.

The missing protein could, the experts say, be obtained from peas and beans, nuts and seeds, and a variety of other vegetables. That might be feasible (although it is unlikely to be popular), but the planners' models do not, for some reason, include the possibility of eating animals other than mammals and birds (mussels, clams, oysters and their kin included), although they could certainly fill some of the gaps as they have before. The idea deserves consideration.

Engels, who wrote with such passion on the deficiencies of diet, of housing and of health among the workers of nineteenth-century Manchester, would have been startled by such figures, and by the fact that today many Mancunians and their fellows choose not to consume what he saw as a necessary minimum for a working household. His insistence on meat every day for lunch, with bacon and cheese for supper, sounds positively unhealthy – not to say rather expensive – today.

His own tastes were more epicurean than those he recommended for the proletariat. When not fomenting revolution, he enjoyed riding to hounds with the Marquess of Grosvenor (whose descendants became the Dukes of Westminster and proprietors of the Mayfair estate that lay close to Marx's home in Soho). At the table, he was famed for his lobster salad, and in 1890 he wrote to Laura Lafargue, Marx's daughter, about his seventieth-birthday celebrations: 'We kept it up until half past three in the morning, and drank, besides claret, sixteen bottles of champagne – that morning we had about twelve dozen oysters.'

Friedrich Engels gives new significance to the term 'champagne socialist', but the oyster repast as he entered his eighth decade reminds us that those creatures and their relatives saved us at another critical point in history, many years before we left Africa, at a time when the food chain was close to collapse and *Homo sapiens* faced the risk of starvation. Then, a novel diet helped man to become modern, and in time to invent a new form of society. Perhaps, one day, as they creep back on to our plates in a world that has begun to show alarming similarities to that of ancient times, the molluscs of land and sea will help us to do so again.

CHAPTER 5

THE SCOURGE OF THE SHADOWS

Lord Hunt of King's Heath: 'To ask Her Majesty's Government whether they have any plans to designate a group of healthcare professionals to be accountable and responsible for the prevention of rickets and its complications'.

Lord O'Shaughnessy: 'Given the widespread availability of Vitamin D supplements and clear guidance to health professionals and the public, the Government does not believe there is a need for further strategies to prevent rickets'.

House of Lords Written Question and Reply, November 2017

In 1942, a secret government survey showed that the morale of the British public had reached a low point. Singapore had fallen, Leningrad was under siege, and in North Africa the Allies were in retreat. To combat defeatism, the government put out thousands of propaganda posters. Some bore pictures of handsome houses and villages, or glorious countryside.

Others showed bombed buildings under stormy skies with a juxtaposed image of a modern replacement, bathed in sunshine. All were decorated with the slogan 'Your Britain: Fight for it Now'.

On the insistence of Winston Churchill, just one was kept from the public eye. It featured a modernistic clinic fronting a piece of urban waste. There, upon a pile of rubble, stands a skinny and half-crippled child. The sun sparkles on the centre's painted concrete, but the downcast boy is shrouded in gloom.

That was a step too far for a politician who hoped for a return of the familiar social order once the war was over. Even worse from the Prime Minister's point of view, the structure shown, the Finsbury Health Centre, had been designed by Berthold Lubetkin, a Russian émigré and communist, best remembered today for his Penguin Pool at London Zoo. Just a few months before the poster campaign, he had built an elegant cabinet in concrete, marble and granite to house a bust of Lenin. It was placed in front of the bombed remains of the house occupied by the Soviet politician at the time of his 1905 visit to London. The structure was less than popular with some of the locals, and Lubetkin was obliged to ask the Russian embassy to lay in a store of heads to replace those damaged by vandals.

His health centre, and Lenin's sometime lodgings, were both in the borough of Finsbury, north of the City of London. The area was then poverty-stricken, with – a century after Engels's description of the slum conditions in Victorian Manchester – a third of the population still forced

A haar drifts from the North Sea across Edinburgh as its inhabitants shiver.

A temperature inversion in the Pyrenees – the hillside snails bathe in sunshine while those in the valley see only clouds.

Author's collection

Taken in the Anza-Borrego desert in 1971, this photo shows (from left to right) Theodos
Dobzhansky, Steve Bryant, Dick Lewontin, the author (with moustache) and Tim Prou

Our Wild Life Photography/Alamy Stock Photo

Cepaea nemoralis snails showing genetic variation in shell colour and number of bands.

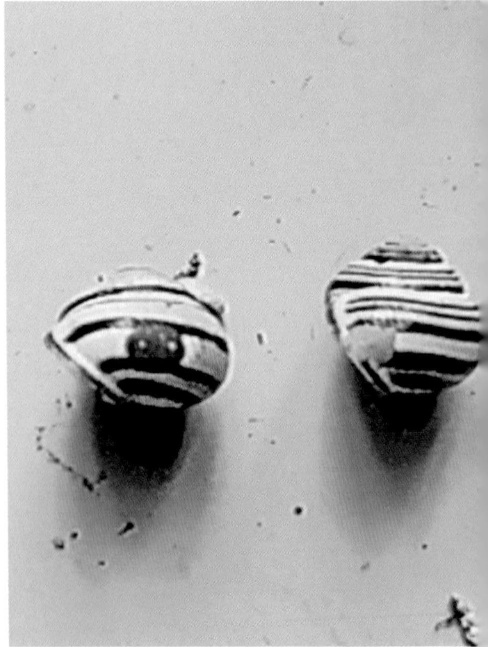

Cepaea snails marked with green fading paint af
three months in the wild; the animal on the left
been less exposed to sunshine than that on the

ıe Resurrection Window in Lambeth Palace: between the sleeping soldiers is a trinity of snails, seen as symbols of rebirth as Jesus rises from the tomb.

l's-eye view of the sun through the canopy of a wood, as seen by a fish-eye camera.

Mediterranean snails climb an olive branch to get away from the heat on the surface.

Shutterstock

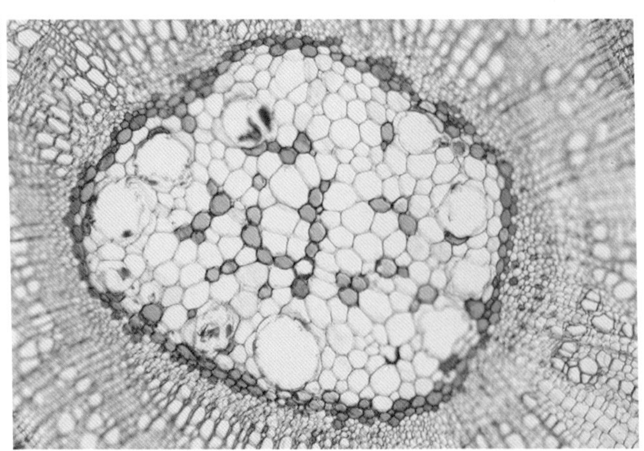

Plant stem section showing central core of vessels that transport water from roots to leaves.

Paul D Stewart/Nature Picture Library

Steller's sea cow, last seen in 1768, just twenty-seven years after it was discovered.

Ann Phillips/Alamy Stock Photo

A midden of abalone shel
in South Africa, showing t
importance of seafood in
Stone Age diet.

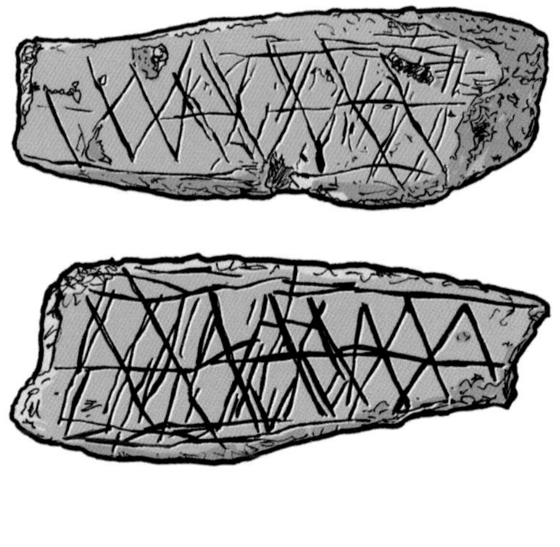

'3,000-year-old piece of ochre marked with lines, from the Blombos Cave in South Africa, then close to the shoreline. The oldest known piece of art.

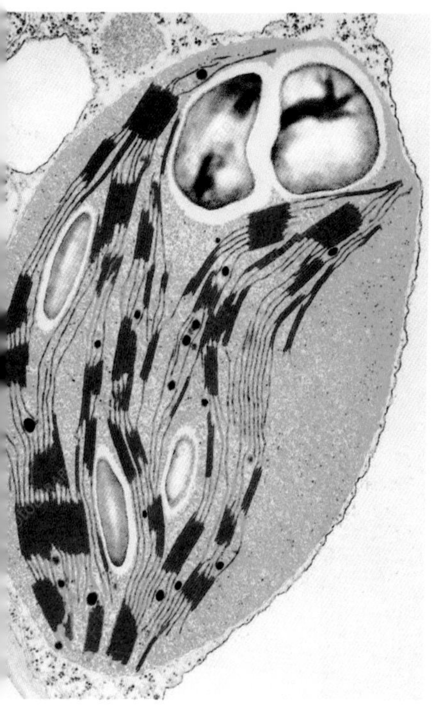

ectron micrograph of a chloroplast, ng the stacks of chlorophyll-containing that pick up solar energy.

Photographic photosynthesis: a portrait made by shining white light through a photographic negative on to germinating grass seeds in a darkened room.

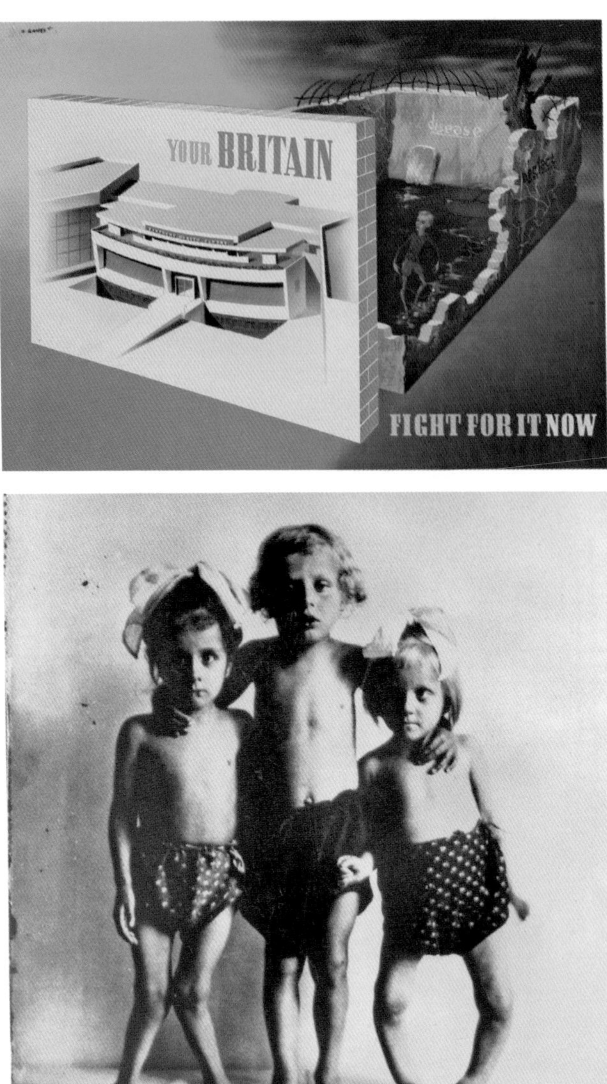

Imperial War Museum

The wartime propaganda poster – banned by Winston Churchill – showing the Finsbury Health Centre.

Wellcome Collection

Children with rickets: soft an
flexible bones, caused by a shortage of Vitamin D.

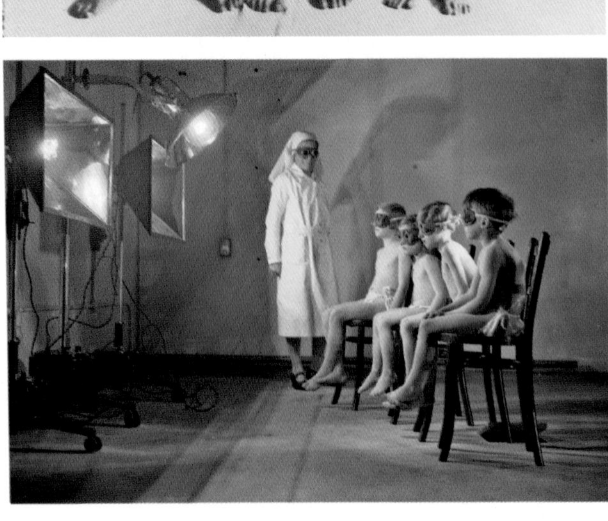

Fox Photos/Stringer/Getty Images

Post-war children bathed in ultraviolet light to synthesise Vitamin D and prevent ricke

John Norman/Alamy Stock Photo

A replica of ust's cork-lined bedroom and riting chamber, where he spent last three years of his life and completed *À la cherche du Temps Perdu*.

Mer de Glace in 2007 and 2016.

NASA

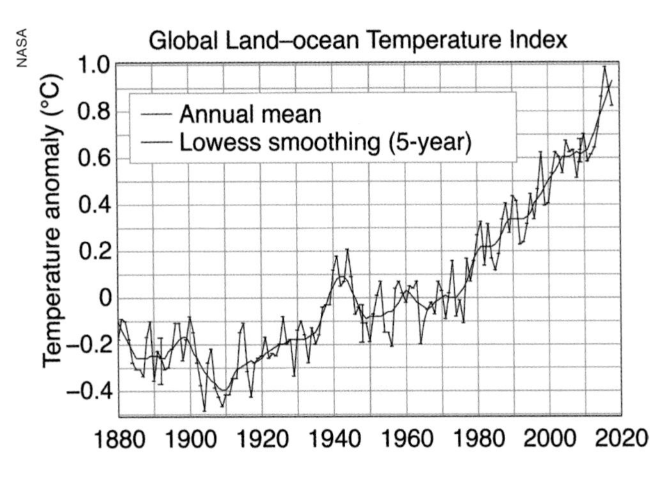

The US National Aeronautic and Space Administration's graph of changes in surface temperature from 1880 to to

The Holme Fen Post – driven into a drained lake bed in 1851 with only the tip showing, now with four metres visible as evidence of loss of carbon from the newly exposed soil.

NASA/GSFC/METI/ERSDAC/JAROS and U.S./Japan ASTER Science Team

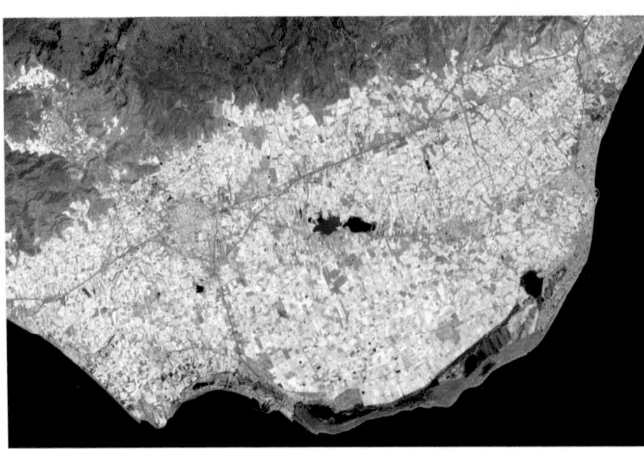

The *Costa del Plastico* at th southern tip of Spain. Fiv hundred square kilometre of plastic greenhouses refl light back into space and the local climate.

to live more than two to a room, and with hunger and disease widespread. In 1934 Finsbury elected a radical Labour council, which set out to transform the lives of its voters. Health was at the top of the agenda. The driving force was an Indian immigrant and doctor, Chuni Lal Katial, later elected Mayor of Finsbury (and the first Asian to hold such a post). He did his best to live up to Lubetkin's motto that 'Nothing is too good for ordinary people', and with his plans for the Finsbury clinic he succeeded. The 1938 structure was the foundation stone of what a decade later became the National Health Service.

Churchill banned the poster because to his eyes it seemed 'a disgraceful libel on the conditions prevailing in Great Britain before the war'. It was not. The child had rickets, a condition once referred to across the world as the English disease, and known by then to be due to a shortage of Vitamin D. That compound has many talents, among them the control of bone formation and the ability to set the level of calcium (essential for bone health) absorbed from the gut. The vitamin is found in cod liver oil, and can also be made through the action of sunlight on naked skin. A shortage, we now know, leads to a variety of conditions beyond the skeleton, infectious diseases and perhaps even cancer included. Others among the borough's infants had tuberculosis, which was widespread in Britain's inner cities and damages bones and joints as much as it does the lungs. The centre set out to fight them both, with sunshine high on its agenda.

Rickets became a scourge at the time of the Industrial

Revolution and was still widespread in the years before the Second World War. The illness was by then unique among non-infectious diseases, as a condition for which a cheap and effective means of prevention was available but for most people was never used.

The malady was first described in 1645 by the physician David Whistler in an eight-page doctoral thesis to the University of Leiden entitled *De morbo puerile anglorum, quem patrio idiomatic indigene vocant the Ricket*. It was 'a peculiar and domestic scourge to our English infants'. Their bones become 'flexible like wax that is rather liquid, so that the flabby and toneless legs scarcely sustain the weight of the superimposed body ... the back, through the bending of the spine, projects hump fashion in the lumbar region'. Affected children 'are too feeble to sit up, let alone to stand, when the disease is increasing'. Whistler saw it as a new illness and blamed it on alcohol taken during pregnancy. He christened it 'Paedosplanchosteocaces' but the term did not catch on.

Rickets can cripple, and some women who suffer from it find it hard to give birth because their pelvis is deformed. As well as their skeletal problems, children may have bad teeth, and the irritability that comes from constant pain. The adult form, osteomalacia, causes thin, soft and often painful bones. Extreme cases may suffer potentially lethal convulsions, seizures, and heart failure caused by a shortage of calcium in the body, essential as it is to nerve transmission.

For many years, nobody knew what lay behind the condition. John Snow, famous for his discovery that cholera is carried by polluted water, insisted that the fraudulent addition of alum (a compound of potassium, aluminium and sulphur) to flour was responsible. Others blamed it on poor heredity, on a lack of breast-feeding (with milk replaced by 'pap', a mixture of bread and water), on syphilis and even on dirt.

All those ideas were wrong. The first hint of the real explanation emerged from north of the border. Edinburgh's haar and smoke had long been notorious and led Robert Southey after his 1819 visit to comment that: 'You might smoke bacon by hanging it out of the window'. Sixty-five years later a local physician, Theobald Palm, who had seen many cases among his own patients, went as a missionary to Japan. He noted that the illness was unknown there, and speculated that its absence was due to the country's benign climate. The people of his home town – Auld Reekie as it was waggishly called after its murk – lived, he wrote, 'under a perpetual pall of smoke and where high houses cut off from narrow streets a large proportion of the rays which struggle through the gloom ... the direct opposites of the conditions prevailing in the "Land of the Rising Sun" ... the most rational treatment would be the systematic use of sun-baths'.

The illness is due not to strong drink or adulterated bread, but to a shortage of a vital food component. Vitamin D has now revealed itself as central to the body's economy. It was identified a century ago as a substance essential to the control of calcium levels in blood and bones, but is now known to influence health and happiness in many other ways. Oily fish, eggs and mushrooms all contain some of it. Fish, in particular, are a valuable source, for a single serving of salmon will provide two-thirds of daily needs (a large egg donates only a twentieth of that). A single oyster, too, has half the recommended daily dose of the compound, providing another reason why their return to the British plate should be encouraged. However, the diet of most citizens in the nineteenth and mid-twentieth centuries – and that of many of their descendants today – provided far too little to satisfy their needs. The main natural source is bright sunshine on exposed skin. Both commodities are hard to find in the British winter (and in some places the summer as well), and were even rarer in the nation's smoky cities before the air was cleaned up after London's Great – and lethal – Smog of 1952.

Sunbaths have long been assumed to be good for health. The ancient Egyptians prayed to the sun god Ra to restore lost hair (the results are not recorded), while in the nineteenth century the 'sun cure' was a popular treatment and preventative against a variety of ailments (some of which actually responded to it). Sunlight has now emerged as a potential remedy for bone disease, for tuberculosis, for various cancers and for a number of other conditions, both mental and physical. A life kept shrouded from it, either because it is screened from view by clouds or by smoke, or – as in this age of electronic entertainment – by the simple choice to stay indoors, is even more harmful than the founders of the Finsbury Centre had imagined.

As peace returned to Europe, a newly elected British government, in a radical exercise in public health, set out to conquer both rickets and tuberculosis. Within a decade the English sickness had almost been driven from its native land

and, with the introduction of antibiotics, tuberculosis too was pushed back. However, political squabbles and changes in the nation's way of life meant that the impetus was lost, and both have begun to return.

Elsewhere, neither ever went away. In Mongolia at the turn of the present century seven out of ten children had the bone disease. Perhaps even worse, more than two billion men, women and children – almost a third of the world's population – still carry the agent of tuberculosis. For most, the bacterium stays latent and causes no harm, but two million people a year still die from the infection.

In 2016 an International Commission reported that rickets and its adult equivalent are 'fully preventable disorders that are on the rise worldwide and should be regarded as a global epidemic. We advocate for eradication through Vitamin D supplementation of all infants, pregnant women, and individuals from high-risk groups and the implementation of international food fortification programs.' In many places that advice has been followed, but in Britain the task was neglected until not long ago.

Tuberculosis, which had been endemic in the eighteenth century and spread with the Industrial Revolution, began to fade away as houses grew less crowded, food improved, and at last as antibiotics came into use. In the last few decades, however, its incidence in the developed world has shot up, in part because of the AIDS epidemic, which made those with the virus more susceptible to all forms of infection, but also through immigration, with around three-quarters of British cases in people born outside these islands. It reached a peak in England in 2011, with more than seven thousand deaths. Most were in London (a city with more than a third of its inhabitants born overseas), which remains its European capital.

Rickets, too, has had a turbulent career, with wild swings in prevalence. As was the case for tuberculosis, all, or almost all, have been driven by shifts in social conditions. Genes, fossils and the records of the past all show how changes in behaviour can alter the incidence, and the targets, of that disease, and of many others.

DNA brings evidence from before records began. Some eighty thousand years ago a small group of men and women abandoned the sunlight of Africa and set off northwards. Under those cloudy skies they faced a shortage of Vitamin D as the dark melanin pigment in their skin cut down the sunlight. They had to adapt, or perish. Natural selection – inherited differences in survival and reproductive success – did the job, and, as it lightened skins, did it fast. Its ability came because rickets is a disease of the young, and juvenile deaths are key to evolution because those who perish in childhood will never pass on their genes.

Several other illnesses have been blamed on a shortage of the vitamin. From cancer to multiple sclerosis and heart disease, they tend to affect older people. Many of their victims will already have had at least some children and have hence passed the test of natural selection, albeit sometimes with a low mark. That may have reduced the sensitivity of such ailments to the vitamin, but to a lesser degree than for the diseases of youth with which it is most associated. That may be why both rickets and tuberculosis are so sensitive to low levels of the stuff, and why even a small supplement can lead to improvement. The ailments of adulthood, less exposed to the Darwinian machinery as they have been, may need more of the compound to keep them at bay. If so, the blood levels now recommended by most experts might be far too low to have a positive effect on such conditions.

Black skin evolved to keep out ultraviolet light. The chimpanzee is pale beneath its fur, but when the first members of the lineage that became *Homo sapiens* came down from the trees, they began to spend more time in the open. To cope with the heat on the surface, they lost most of their hair, which meant that their naked flesh was exposed to bright sunlight. That destroys another vitamin, folic acid, a shortage of which interferes with the development of a baby's spinal cord, and is hence a powerful agent of selection. A million years and more ago our ancestors evolved dark skin to deal with the problem.

In Africa enough ultraviolet still got through to protect against rickets, but the emigrants faced vitamin deficiency as they moved towards cloudier skies. Only those with less pigment than average could generate enough of the compound to keep their skeletons in good shape. They survived and passed on their heritage more often than did their swarthier fellows, and, as the generations succeeded, European faces took on a pallid hue (with those of today's Scots the pastiest of all). Across the globe, such complexions are rare, for just one person in fifty, worldwide, is blonde, and before Columbus began to complicate matters every one of them lived within a couple of thousand kilometres of Copenhagen. Redheads are rarer still, and yet more localised. Many among the Celts of the far north and west of these islands have lost even the ability to tan, so effective has selection by shortage of sunlight been.

Post-war migration means that around one person in seven in Britain now has dark skin, and the incidence of the bone disorder in that population is more than ten times higher than in their fellow citizens, among whom eight people in every hundred thousand are known to face the problem. The true incidence in both groups may be considerably higher, for post-mortem examination of children's bones shows that in some places more than half have early signs of the illness that could not be detected on an X-ray. When assessed in this way, its incidence is higher on these shores than anywhere else in Europe, so that the condition is still, as it long has been, an English – or British – illness.

The transformation from black to white began no more than twenty millennia ago. Around a dozen genes were involved. Fossil DNA from the Ukraine shows that five thousand years before the present most south-east Europeans were much darker than they are today. The remains of Cheddar Man, who lived in south-west England five millennia earlier than that, show that he too had dark skin, in a hint that the British move towards pallor had by then scarcely begun.

The earliest direct evidence of bone disease lies in the misshapen remains of a woman in her twenties, a contemporary of the fossil Ukrainians, who lived on the Hebridean island of Tiree (*Tir Iodh*, the 'Land of Corn'). Its people grew barley and herded cattle and sheep. The seas around teem with fish, and Tiree is the sunniest place in Scotland. It seems odd that she was short of the vitamin, but perhaps, as often happens, the island's farmers ignored the bounty of the ocean and she paid the price.

Skeletons show that rickets was common across the Roman Empire three thousand years later, with about one child in twenty affected and with Britain, even in those days, having the highest incidence of all. Later, in Greenland, the disdain of one culture for the cuisine of another had the same effect. Erik the Red (named for his hair colour), with four hundred settlers, landed there in 986. Like the people of Tiree, the Vikings grew cereals and grazed livestock, and had little interest in the habits of the local seafarers, who ate vitamin-rich fish every day. They arrived in a relatively warm and sunny period, but quite soon the weather began to change, their crops failed, their cattle died, and soon they were obliged to hunt seals for their meat. By the mid-1500s, soon after the onset of the period known as 'the Little Ice Age', the Greenland colony was extinct.

Skeletons from its final years showed that many children died young, with signs of bone disease. Adults, too, had curved spines, deformed arms and narrow hips. In that sunless and misnamed land nobody could make enough Vitamin D to stay healthy even at the height of summer.

In Renaissance Italy, the ailment turned its attentions to the rich. The remains of sixteenth- and seventeenth-century Medici children from the Basilica of San Lorenzo in Pisa show signs of its presence, no doubt because they were wrapped in swaddling clothes, kept indoors to preserve their pale complexions, and fed on breast milk – a poor source – for far longer than were the infants of the peasantry, who played in the sun and were weaned as soon as possible. David Whistler, the doctor who described the syndrome, wrote of his own country in 1645 that: 'The disease is most frequent in the ranks of the highest citizens.' He may have been right. Charles the First (who died of an unrelated skeletal condition) had a posture problem, and the bones of his daughter Princess Elizabeth hint that she too suffered from the illness. One of the treatments recommended in his day included warming the abdomen in the sun, which might even have done some good.

In London at that time, the Bills of Mortality, the statistics on causes of death, also saw rickets as rare, and in 1634, when the word first appears in their pages, just fourteen of its thousands of deaths were ascribed to it. Physicians referred to it as a new arrival, due, they thought, to the city's damp climate. In fact the clouds, rather than the rain, were to blame.

Those clouds became filthier each year, and the capital's inhabitants began to pay the price. John Evelyn in his 1661 pamphlet *Fumifugium* ('The Inconveniencie of the Aer and Smoak of London Dissipated') railed against its air: 'Ecclipsed with such a cloud of Sulphure, as the Sun itself, which gives day to all the World besides, is hardly able to penetrate and impart it here; and the weary Traveller, at many Miles distance, sooner smells than sees the City to which he repairs.' He blamed the problem on the use of sea coal from

Newcastle by brewers, soap makers and the like, and pleaded, to no avail, that such industries be moved out of the city into suburbs filled with sweet-scented flowers. A quarter of a century after his diatribe the incidence of rickets had begun to shoot up.

By the middle of the nineteenth century, London could boast of two million citizens, and the English sickness was rife. In Manchester, conditions were even worse. In both places, coal use had gone up by ten times since John Evelyn's day and in each the air was black with smoke.

Nineteenth-century skeletons from the Cross Bones Cemetery in London's Southwark – a district now filled with bearded hipsters but then a ruined and filthy slum – showed that almost half the children suffered from the condition, five times as many as in leafy Chelsea, four kilometres away. Even so, the authorities took little notice, in part because the bone disease was at the time not regarded as a condition that should be recorded on death certificates.

In those years tuberculosis (which was documented in this way) was also a scourge, albeit one surrounded, in fiction at least, by romantic imagination. It had a host of names – consumption, phthisis, scrofula, the King's evil, and more – each of which was assumed to be a distinct disease. Their cause was unknown and the treatments were based on fantasy. In England, to be touched by the King (or even to wear a necklace that bore a coin with his image) was found to be effective, but in Scotland the hand of an executed criminal was preferred. Failing that, a broth made from leeches was often successful.

The condition is, we now know, a single syndrome caused by a bacillus called *Mycobacterium tuberculosis.* It attacks many parts of the body, from the bones and joints to the lungs and muscles, and in some cities – Edinburgh and London included – once killed as many people as did all other infections put together (Boswell's wife and his oldest daughter each died of it, as did thousands of less well-remembered Scots).

In 1848, doctors at the Brompton Hospital for Consumption and Diseases of the Chest began to try new treatments. They gave more than five hundred people with the illness a dose of cod liver oil ('patients take it in general without repugnance'). For one in five, the disease was stopped in its tracks: 'the cough was mitigated, the night-sweats ceased, the pulse became slower and of better volume; and the appetite, flesh and strength were gradually improved'. Patients so treated survived twice as well as those given vegetable oil instead. Nobody had any idea why the fish extract was so effective.

Sunlight, too, began to play a part. A sunbath had already been found to help the skin disease lupus vulgaris, a disfiguring facial condition caused by the agent of tuberculosis, and sometimes a side-effect of the lung disease itself. In the late nineteenth century, as many as one person in fifty was affected. Soon, an arc lamp was used instead, with real success. The treatment was invented by the Dane Niels Finsen, who received a Nobel Prize for his work.

It then emerged that ultraviolet light also helped other skin conditions such as psoriasis, eczema and acne. Some doctors even began to open up tuberculous cavities around joints and irradiate them to fight the bacillus. Within a few years such 'electric light cabinet baths' were in use against lung disease, heart problems, and even against the gangrenous wounds of the First World War.

Sunlight came into its own when it was discovered that when enough was available it could also help against the internal forms of tuberculosis. The Swiss physician Auguste Rollier began to worry about the condition when his fiancée contracted it, and, inspired by the work of Finsen, set up the first of many 'heliotherapy' clinics in the Alps. His patients followed a strict regime. For the first few days, they lay covered in a white sheet, with just their feet visible. Over a week or so, the whole body was exposed, and soon the patient was tanned and in time perhaps cured. One medical visitor was impressed to see patients 'as black as negroes, with considerable and healed tubercular wounds'. Rollier became a proselytiser for sunshine, which he felt also produced '*joie de vivre*' and a vigorous work ethic. His treatment (and a few of the joys of life) plays a part in the plot of Thomas Mann's novel *The Magic Mountain*, in which the wealthy characters set up a 'Half-Lung Club' in an Alpine sanatorium as they suffer elegant decline accompanied by intellectual small talk, in one of which they consider – appropriately, given their condition – whether life is no more than an infectious disease of matter.

Another exponent of the sunlight cure was Robert Louis Stevenson, who fell ill while on honeymoon in California. He noted in a letter to his mother that:

> This life takes up all my time and strength. By the time I have had my two sunbaths and my two rubs down with oil . . . I have no stomach for more. I am truly better; I am allowed to do nothing; never leave our little platform in the canyon, nor do a stroke of work; that and sunbaths and oil are, I think, doing me great good.

Britain's doctors soon embraced the treatment. The nation's pioneer of sun therapy, Sir Henry Gauvain, was a surgeon in the Cripples' Hospital at Alton in Hampshire in the years around the First World War, and an expert on the repair of joints damaged by tuberculosis. He had a more robust attitude than did his Swiss fellows. For him, the British climate was ideal, as the sun came and went at short intervals rather than in the tedious day-long glare that visitors to the Alps were forced to endure. To make his cure more enticing he combined sunbaths with dips in an icy ocean. In settings less grand than those of Davos (or Stevenson's Napa Valley), large numbers of British children were sent to Sea Side Hospitals in the hope that the open air would improve their mental and physical health.

As was the case for cod liver oil, nobody had any real idea why sunshine had such positive effects. Various theories surrounded the practice. Florence Nightingale, for example, liked bright light because it dispelled 'miasma' – 'bad air' – the supposed source of all infection, those of the lungs included. She, like many others, was wrong.

The skin form of tuberculosis responded to Finsen's lamp because the bacterium is very sensitive to the lethal effects of ultraviolet light when exposed directly to it. Transmission of the lung infection is by particles coughed out by a carrier, and many treatment clinics now use ultraviolet lights to reduce the risk of cross-infection.

All that seems simple enough, but how to explain the beneficial effects of sunlight on infected lungs, or joints, which never see it? The explanation emerged from work on the biology of rickets.

The value of shark liver oil in its treatment had been discussed by an Edinburgh doctor as long ago as the 1720s, but was not taken up for many years. In the late nineteenth century it was noted that rachitic lion cubs in London Zoo, deprived of ultraviolet because they were kept behind glass, could be cured with an extract of fish liver. Soon, its use became widespread in Scandinavia and elsewhere, although again nobody had any idea how it worked. The key, we now know, was Vitamin D, made in the liver by fish and in the skin by men and women when exposed to sunlight, and transported to almost every cell in the body.

An early hint that the positive effects of ultraviolet light were due to a substance carried in the blood emerged at the time of the British blockade of Germany in the First World War, which led to widespread hunger. As many as half of the nation's orphanage children developed the English disease. A local doctor made the crucial observation that exposure of just one arm to ultraviolet rays improved its symptoms throughout the body.

German scientists at that time insisted that a careful balance of protein, fat, starch, sugars and minerals alone would make an ideal diet. Such confidence was unjustified, for they overlooked the importance of vitamins. Hints of their role had come years earlier, when an Edinburgh-based naval doctor, James Lind, in one of the first controlled clinical trials, found that citrus fruits protected mariners against scurvy during long voyages. Vitamin C itself was discovered in 1912.

Seven years later, the English physician Sir Edward Mellanby became interested in the possibility that fish oil might contain such a substance. He noted that skeletal problems were more common in Scotland than in England, and that the Scots diet was, as it still is, less nutritious than that of their southern neighbours. North Britons of that era ate large quantities of porridge, a dish based on oats (in Dr Johnson's much-quoted definition, 'a grain, which in England is generally given to horses, but in Scotland supports the people'). Mellanby tested the effect of such a diet on dogs, which in a further simulation of the northern way of life he kept out of the sun. All developed severe bone problems that were reversed when the animals were given cod liver oil. At first the effect was assumed to be due to the presence of Vitamin A, identified in butter-fat a few years earlier, but when this was destroyed by bubbling oxygen through the oil, the liquid still cured his animals. Vitamin D had been discovered.

Its importance became obvious in macabre experiments in which rachitic rats were fed skin from their fellows, for only if the donors had been exposed to ultraviolet did the animals' bones return to normal. Four years after Mellanby came the unexpected report that rats moved to empty jars which had been irradiated grew just as well as animals with direct exposure to ultraviolet. In fact there was a residue of sawdust in the jars, and the rays activated that to make Vitamin D. It could do the same for other foodstuffs, none of which helped people with the disease until they were irradiated. Even rickety rats housed in cages with healthy animals were cured, for they ate the enriched droppings of their neighbours. Irradiation is now the standard method used to supplement foods (and even beer) with the chemical.

Mellanby's discovery caused the campaign for sunlight to be taken up with renewed enthusiasm. It was given impetus by the Edinburgh physician and eugenicist Caleb Saleeby, who identified what he called the 'diseases of darkness', rickets, tuberculosis, drunkenness and depression. In 1924 he founded the Sunlight League, which promoted what he called 'helio-hygiene', which involved the maximum exposure to 'nature's universal disinfectant, stimulant and tonic'. It soon spawned a spin-off group, the Men's Dress Reform Party. That had his fellow Scot the medical missionary and rickets pioneer Theobald Palm as its first president, and claimed that: 'Most members will be for shorts; a few for the kilt; all hate trousers ... *but the villain of the piece is the collar stud.*' The organisation was much mocked, not least by George Orwell, who himself died of tuberculosis and who was rather less of a liberal than he is often painted. In *The Road to Wigan Pier* he sneered at those who supported such namby-pamby behaviour. They were fruit-juice drinkers, nudists, sandal-wearers, sex maniacs, Quakers, 'Nature Cure' quacks, pacifists, and – perhaps worst of all – feminists; but in spite of his scorn a glance at a summertime street today shows that the campaign to persuade the British to cast off their collar studs has had some success, even if kilts are still rather a rarity except at the time of the Edinburgh Festival. Germany went further in casting off its clouts, for in the 1930s nudism became popular, as it still is; while in Russia, Stalin was one of the many keen users of sun-ray lamps and vitamin treatments.

As both Stalin and the Sunlight League had been quick to notice, the means to purge the bone disease were simple and ready for use, but it took another quarter-century before the stage was set for a real assault.

Winston Churchill had been outraged by the poster that hinted at the deprivation faced by Britain's urban children. He was supported by the royal physician Baron Horder, who had assured the public not long before that rickets was 'fast dying out', although in his day the majority of inner-London children had signs of it. Horder, like many of his successors, was an opponent of the 'nanny state', and in his view every citizen had the right to go to hell in his or her own way, whatever the cost in human misery.

As the war ended, the nation's voters rejected that philosophy. In the election of that year Churchill's hopes for a return to what he saw as the natural order were dashed. Among the first goals set by the new Labour government was to eliminate rickets, which, as one MP said, had 'ceased to be a medical disease and had become a political one', and at least to try to do the same for tuberculosis. For a period, at least, both schemes succeeded.

I was born a year before the Attlee administration took

office, and grew up exposed to the full armoury of state-sponsored optimism. Vitamin D and calcium were added to flour and to margarine, while pregnant women, nursing mothers and young children were provided with supplements. Dried dairy products and free school milk did a lot to build up blood calcium. Some of my earliest memories are of being force-fed cod liver oil at the age of five or so as I clustered naked with my female cousins around an ultraviolet light (I was, I seem to recall, somewhat baffled by the physical differences on view). The sole downside was the stench of stale milk that pervaded my schooldays.

In 1954 – a year in which my generation was hailed as 'Young Elizabethans' – victory over the bone disease was announced. The statement was described by the Labour politician and founder of the Socialist Health Association, Dame Edith Summerskill, as the 'most spectacular change which has taken place in this country'.

She may have been right, but the effort to improve the health of the nation soon ran out of steam. A change of government led to the end of subsidies. Its leaders used a (quite unfounded) claim that many children suffered from an overdose of the vitamin as an excuse to abandon the programme, but that claim turned on political expediency rather than medical evidence. The new rulers decided that only the poor, the pregnant and those with small infants deserved help. Dame Edith warned that: 'To make ... the welfare foods a luxury is bound slowly to undermine the fine achievements in the field of preventive medicine ... It will be an insidious process, and only the social historians will be able to point to the folly of this action.' The fight against the English disease in its native land had become an ideological football rather than an exercise in disease control. The selfishness and stupidity of that decision took less than a decade to emerge.

The return of rickets came to the attention of doctors in the early 1960s when they noted an increase among Scottish children. That concern was soon submerged by the publicity about its appearance among Asian immigrants. It was without doubt present in some of those families, but there grew up a sense that they should face criticism rather than sympathy: as one civil servant said, the condition was 'a sign of social regression'. As had been the case when tuberculosis was blamed on the arrival of Jews, the appearance of an 'alien disease' was used to argue against immigration.

Politicians found the rise in incidence such a sensitive issue that it was not discussed in Parliament until it came up in a 1971 debate about the decision by Margaret Thatcher, then Education Secretary, to abolish the provision of milk in schools to children over seven. Free milk had been on the menu since the start of the health campaign, but had been removed from secondary schools by the previous, Labour, administration. In the debate, the Member for Bedwellty, Neil Kinnock, claimed: 'Now this atavistic Government have moved back to the priorities of a bygone age ... This is a barbarian Bill which is the product of a barbarian mind.' Mrs Thatcher in response insisted that a survey in Birmingham had shown that only immigrant children (as her colleagues tended to call British-born children of the new arrivals) showed signs of the disease, but ignored the study's main conclusion, that one in five of the city's young people, migrants or not, had low levels of the vitamin. Kinnock notwithstanding, the policy went ahead and the issue faded from the public mind.

That was a mistake, for a shortage of Vitamin D and of calcium was then, as it still is, a real problem. Today, just one in three British teenagers consumes the three or more portions recommended for calcium-rich dairy foods each day, while twice that proportion have a can of soda instead. Strict vegetarians take in almost none of the vitamin. Digestive upsets can reduce the amount absorbed, while some inherited variants hinder its action. However, the greatest danger comes from a change in Britain's attitude to the outside world, most of all among the young. Almost unaware, a whole generation has slipped into a new relationship with daylight. Today's vitamin deficiency is due as much to social revolution as to political failure.

In my own schooldays, free of electronics as they were, we were driven outdoors for at least an hour a day even in icy weather. We assumed that this was to give our teachers a chance to smoke their pipes and discuss the inadequacies of their pupils, but in reality it came from the then widespread belief in the benefits of the open air. My suburban Merseyside school backed on to a large field with, to improve matters, a piece of woodland in which we could disappear from adult supervision. Both are still there, but that is unusual, for hundreds of hectares of school sports fields have been sold off. They are still being shed at a rate of a hectare a week.

Millions of young people have retreated into an endless gloaming as they peer into a computer, a games console, a mobile phone or a television screen. By the time they enter their teens, nine out of ten British children have their own phone, and almost as many their own tablet. Even worse, perhaps, more than half of all seven- to eleven-year-olds have a television in their bedroom (which for many stays on even as they sleep), while one in three fifteen-year-olds spends at least six hours a day online. Britain is second only to Chile in that regard.

This island's children are now among the least active in the world. The United Nations Standard Minimum Rules for the Treatment of Prisoners insist that every inmate should spend at least one hour a day in the open air. Just one in six British teenagers lives up to that injunction. There has been a decrease of a fifth in muscle strength as measured by hand grip and the ability to do sit-ups among ten-year-olds over the past decade, evidence perhaps of a childhood deprived of outdoor exercise. The nation's pensioners also shun the daylight, with an average of just half an hour a day out of the house.

All this can lead to a dearth of Vitamin D. It has been exacerbated by another social change. The compound is soluble in fat, which means that overweight people tend to have too little in the blood, for it has been soaked up by their fatty tissue. The incidence of childhood obesity has risen by four times since my schooldays, and the bones as much as the flesh of those who suffer it may pay the price.

One widespread rule of thumb, based on the compound's

association with bone disease, is that a level in the blood of less than twelve nanograms – twelve-billionths of a gram – in every millilitre represents a real deficiency. For most authorities any level over twelve nanograms is acceptable, although some call for thirty.

A recent survey of tens of thousands of people across Europe showed that around one in eight had less than the recommended level in their blood over the year as a whole, a figure that reached its low point in winter. Some countries call for a level of forty nanograms instead, and by that criterion the proportion who fail to reach the recommended level rose to almost half. Real evangelists claim that seventy-five nanograms are needed, and on that counsel few people reach the target. In a survey of Holland, Germany, Poland, Spain and the United Kingdom, our homeland came bottom of the list, with eight times the proportion of vitamin-deficient people as in Holland, at the top. One in three British infants is short of the crucial compound, while on the global scale, the problem is even more widespread, for nine in ten Iranian infants face the problem, as do half of those in Turkey.

Decisions as to how much time must be spent outside to top up the substance are complicated by place, by season, and by time of day. Every European is at some risk. From Oslo to the pole, in winter there is no usable ultraviolet at all. North of a line drawn through Birmingham nobody can synthesise enough from natural light alone over the year as a whole to avoid shortage, and even Gibraltar has too little sunlight to generate sufficient Vitamin D in the middle of winter.

A general guide is that a white person of average hue in a

southern English June who exposes his or her bare arms, legs and face for fifteen minutes a day three or four times a week in summer will make enough to stay healthy for that season at least. Given that the average Briton now walks for less than ten minutes a day, many fail that test. In winter a low sun angle together with short and cloudy days mean that no citizen, even one who spends most of his or her time outdoors, can reach the threshold. One useful way to check is to look at one's own shadow. If it is taller than the person who casts it, too little ultraviolet is available to be of any use.

That seems simple enough, but how much sun do people really experience? Those who take part in health surveys tend to lie to the interviewer (and to themselves) about how much they drink, smoke or eat, and the same is true when asked how much time they spend indoors.

The quest for the truth led to an unexpected overlap between the study of human behaviour in sunshine and that of snails. The ultraviolet sensor I developed long ago is based on a stable yellow paint mixed with a blue dye that breaks down in sunlight. That green mixture fades back to yellow at a rate that depends on exposure. Albinism – a lack of skin pigment – is common in parts of southern Africa, and those with the condition are advised to stay out of the sun because they face a risk of skin cancer a thousand times greater than average. Some years ago I spent a few months at the University of Botswana. There I discussed with a colleague the idea that we could make caps out of yellow cloth soaked in the unstable dye to use the change back from green to yellow to measure how much time children actually spend in sunshine, rather than believing what they tell their parents. The plan went nowhere, for it failed its ethics exam. The authorities told us that we would have to tell our subjects what the caps were for. That meant, we were sure, that the kids would just take them off as soon as they were out of sight, and we abandoned the idea.

Now a method related to (but perhaps less convenient than) my own is used to ask the same question in the UK. It utilises a plastic called polysulphone that breaks down when exposed to ultraviolet.

Strips of this material were attached to adult Mancunians for a few days in each of the four seasons. The sun is at its highest in the two hours around noon, but even in summer the sensors showed that each subject went outside at that time, on average, for no more than nine minutes on weekdays and twice that at the weekend. From ten in the morning to three in the afternoon (the only period with enough ultraviolet to make a useful dose), the people of Cottonopolis exposed themselves for just twenty minutes each weekday, and forty on Saturday and Sunday. Their vitamin levels were at their highest in September but, even then, just one in four had the recommended amount. A large majority was unable to reach the threshold in winter. Just one in five white Manchester teenagers has a safe level throughout the year, while almost as many are deficient for at least part of it. A quarter run low even in summer, and almost all find themselves below the limit in the short days of winter. In a reminder of the fate of their ancestors in the days of the Industrial Revolution, many among them have thin bones.

Mancunians of Indian and Pakistani ancestry had an even more distant connection with sunlight. Their darker skins fend off its rays, and their habits do not help. Asian men spend about the same time in the open as do Europeans, but the sensors show that they tend to seek the shade, perhaps as a relic of the habits of their homeland. The amount in their blood goes from rather too low to disastrously so. Just one in ten reaches the recommended level at any time, and many have so little that they face real danger of bone damage, even as adults. For some of their female partners the problem is exacerbated by their thick clothes and by their reluctance to spend time outside (and the vitamin issue is almost never raised by those who argue for, or against, the right to wear a full-face veil). Asian men in Manchester would have to spend four times longer in the sun each day than their European equivalents to reach the recommended level, and their wives and daughters might never attain it. Nine out of ten young Asian women had less than half the necessary amount, while a quarter had almost no Vitamin D at all. One in sixty has signs of rickets, an incidence several times greater than in their European fellows.

Across Britain as a whole, about eight million citizens are short of the stuff, with Afro-Caribbeans at twice the risk of whites, and Asians three times. In winter the citizens of these islands even fall behind the people of Scandinavia, for the latter have long dosed themselves with cod liver oil. There, too, immigrants do much worse than the native-born. In Finland, Kurdish refugees from Iraq are seventy times more liable to have very low levels than are the locals.

At the time of George Harrison's homage to sunshine, Britain saw a little more than one case of the disorder in each hundred thousand, but at the turn of the millennium the figures began to climb. The number of episodes is now five times higher than in his day, and is still rising.

The return of rickets and of tuberculosis has added new urgency to the study of the curative substance itself. It is a hormone rather than a vitamin, for it alters the activity of genes, rather than – as many vitamins do – acting as a catalyst for biochemical reactions. One form is made by plants such as shiitake mushrooms, while another is made in the skin of most mammals (sheep are good at it, and their wool is full of lanolin, a steroid related to cholesterol, so much so that some dietary supplements are made with ultraviolet light directed on to a fresh fleece).

The first step in its career comes when ultraviolet B (which does not penetrate far into the skin) strikes a relative of cholesterol. This makes a precursor that moves to the liver and the kidneys, which modify it further to make the vitamin itself. It enters cells from the blood through specialised gateways which come in a variety of versions that alter the rate of passage. Some show regular patterns across Europe from south to north, perhaps because the northern variants speed up delivery to the target.

Once inside, the chemical alters the activity of the genes behind bone formation and cell division. That is not unexpected, but a dose of the substance given to cells in culture also changes the activity of more than three thousand other genes. Some are involved in brain growth and maturation while others are active in the production of blood cells, in the immune system and in the lungs. What most of the others do is as yet unknown, but as at least half of the four hundred known cell types have receptors on their surface, the compound may have an influence on most parts of the body.

At first sight the bone disease and that of the lung seem unrelated, but their shared response to the same substance was the first of several hints that it works its magic in many ways. Many other illnesses have joined that duo. Some physicians are so impressed by the compound's claimed abilities that they have made an unconscious return to the policy of the Sunlight League, with hope of a new era of good health based on exposure to ultraviolet or, if that is not available, on a convenient pill.

However, as almost everyone knows, sunlight can be harmful, too. The first scientific intimation of its dangers emerged from a survey of American naval personnel in the 1930s. Sailors, with their open-air lives, were found to have high rates of skin cancer, which was blamed on their exposure to ultraviolet. That discovery has been acted on across the globe.

Sir Henry Gauvain, pioneer of British heliotherapy, had already recommended moderation. Sunlight was 'like a good champagne. It invigorates and stimulates; indulged in to excess, it intoxicates and poisons.' That soon became the universal belief about the threats and promises of sunbathing (if not of champagne).

Ultraviolet, like alcohol, packs a considerable punch. Sunburn can be painful and even dangerous. Physical damage to the skin makes the young look old, and too much time in bright light can lead to cataracts and even to blindness. A sudden burn is more dangerous than is long-term moderate exposure. For a pale redhead on a sunny June day at noon, just seventeen minutes of sunlight is enough to harm the skin, while a dark-complexioned southern European is safe for twice as long.

The best-known threat is indeed skin cancer. One form, malignant melanoma, can be lethal. Its incidence matches the level of exposure to ultraviolet light across the globe. Other variants are less dangerous and appear as moles or other blemishes that may need to be removed on safety or cosmetic grounds. As proof of the direct influence of sunshine, in the United States such growths are more common on car drivers' left arms, but in Australia, where they drive on the other side of the road, the opposite is true. The DNA of each is riddled with mutations of the kind caused by ultraviolet.

In countries with fair-skinned populations the incidence of the malignant form of the disease has risen steadily over the past five decades. The latest surveys show that the highest rate of all is in New Zealand, with thirty-six cases per hundred thousand, and Australia just behind it at thirty-five. In Europe, Switzerland, with its keen skiers, is number one, with Scandinavian countries just below that figure. Even within Scandinavia the southernmost population, the Danes, do worse than their neighbours to the north.

Until 2016, Australia had the world's highest incidence of malignant melanoma. The nation has long tried to control it. Its 'Slip, Slop, Slap' campaign enjoins citizens to slip on a shirt, slop on some sunscreen, and slap on a hat (a slogan squawked by Sid the Seagull). Sid is just a local hero, but although similar schemes have been launched across the world the incidence is still going up, and in Britain has done so by half in the past decade, with death rates rising in step.

Convincing as the evidence on melanoma is, and worrisome as the carcinogenic effects of sunlight may be, the US Navy survey came up with another, and quite unexpected, result. Its seamen, with their open-air lives, were less, rather than more, liable than average to suffer from other forms of cancer. Perhaps, came the suggestion, ultraviolet light – dangerous as it was for one form of malignancy – protected against cancers of the prostate, the lung, and more. One physician of the time even suggested that everyone be given enough of those rays to generate non-malignant melanomas (easy to remove should they cause trouble) as a general protection against cancer, just as vaccination protects against infectious disease.

That audacious suggestion seemed to be on the fringes of medicine. The claim became even less credible when it was picked up by private clinics that offered expensive sunlamp treatments even to those with terminal forms of the condition. Hundreds of desperate people were drawn into their web. Bob Marley, who had malignant melanoma, went to a sunlight treatment clinic in Germany in November 1980. He died in Miami seven months later, on his way back to Jamaica.

In spite of such doubtful claims, the evidence that a shortage of sunshine is indeed associated with, and may in part be responsible for, a variety of illnesses, some cancers included, has become more credible. Almost all are – unlike rickets – diseases of older people.

Real enthusiasts claim that, while ultraviolet causes some mortality, on balance the lives that could be saved by experiencing more of it or of its vitamin product outnumber the deaths by a dozen times or even more. Such figures are no doubt overstated, but the evidence of a positive effect of ultraviolet light on general health has become hard to dismiss.

Across the world, a whole range of cancers is less common in sunny places. Leukaemia, together with cancer of the ovary, the bladder and the pancreas, are all more frequent the further one moves away from the equator towards either pole. Prostate cancer affects eighty in every hundred thousand men in Iceland, but fewer than ten in Malaysia. Breast cancer shows a similar decrease in tropical countries in both Old and New Worlds, with an incidence in Malaysia and Laos only an eighth that in Canada and New Zealand. The same pattern shows up within countries. Calabria, in the south of Italy, has a third fewer cancer deaths each year than does Liguria, in the north. One (optimistic) estimate is that in sunny countries there may be two or three extra deaths in every hundred thousand from malignant skin cancer, but forty fewer deaths from all other forms of the disease. Sid the Seagull may, it seems, inadvertently have done more harm than good.

Multiple sclerosis shows similar patterns. The illness comes when the immune system turns upon itself and can lead to blindness and paralysis. Its symptoms appear on average about ten months earlier in life for every ten degrees closer to the pole – the distance between Bordeaux and Edinburgh. Canada has the highest rate, with almost three hundred patients per hundred thousand, and Denmark and Sweden are not far behind, while the condition is almost unknown in Africa and in Central America. Canadians who bear rare inherited variants that lower the amount of blood Vitamin D are at a notably higher risk than are others. In Australia, most inhabitants are of British and Irish descent, but the incidence of multiple sclerosis is around half that in their ancestral islands – a hint that a shortage of ultraviolet lies behind at least some of the cases within our own shores.

Some physicians go further and claim that medical exposure to light of the correct wavelength may slow the progress of the illness, perhaps because it reduces excess activity of the immune system. Vitamin D treatment might also be useful, for it reduces the chances of inflammation, the agent of many of its symptoms, and at least in some patients seems to postpone the advance of the disease.

Juvenile onset diabetes might also be helped in this way. It comes from an inability to secrete the blood-sugar hormone insulin. That too arises from an internal attack by the immune system on the body's own cells. Again, its incidence is U-shaped, with a low point around the equator, and with Scandinavia paired with Tierra del Fuego (in the far south of the Americas) close to either peak, with around twenty times the tropical rate. The illness has become more frequent, and symptoms tend to emerge earlier in life than once they did, perhaps because so many young people see less sunshine than did their parents.

High blood pressure also shows a fit with low levels of sunlight, as does hardening of the arteries. Within most European countries more people suffer from elevated blood pressure and artery damage in winter than in summer, and across Europe, from Finland to Italy, latitude from south to north is a stronger predictor of both conditions than is age, diet, smoking, sex or exercise. As a further hint, in some northern countries people with lots of moles have a lower risk of heart attack, perhaps because such blemishes are signals of past exposure to the sun.

These figures are telling, but say little about the real balance of cost and benefit of a solid dose of sunlight. A huge study in Sweden set out to test the strength of the effects on overall mortality. For fifteen years from 1990, thirty thousand women from their twenties to their sixties were followed. They were divided into three groups: those with low exposure to ultraviolet because they spent most of the time inside, never visited tanning salons, and did not take holidays around the Mediterranean; those who received moderate amounts; and the real heliophiles, who soaked up the vital rays as often as they could.

A positive fit of both forms of skin cancer with sunlight was indeed found, but such rays showed a kinder face when it came to overall survival. At the end of the survey, four in every hundred of the high-ultraviolet group had died, but twice that proportion of those who shunned them had suffered that fate, with much of the effect due to heart disease.

The difference in life expectancy between the low- and high-exposure groups was as great as that between smokers and non-smokers. A more recent survey in Minnesota that looked at death rates of large numbers of people over five years in relation to vitamin levels also found that those with little Vitamin D had more than double the risk of death over the course of the survey than did those with satisfactory amounts.

Those figures are impressive, but a more rigorous approach would be to take a look forwards rather than backwards and to follow the habits, the vitamin levels and the health of a large number of subjects for many years. A survey in Rotterdam of citizens of over fifty-five lasted from 1997 to 2015. As the years passed, almost eight hundred new cases of dementia appeared. Participants with low vitamin levels were significantly more at risk of the condition than were others. The French have set up two such studies, one with around twelve thousand members that began in 1994 and has continued to today, and another with almost thirty times as many that began in 2009. Again, those with low levels of the essential chemical suffered a higher incidence of depression, of prostate cancer and, for smokers, of lung cancer as the survey continued.

The effects of ultraviolet on such a wide variety of conditions have led to new interest in just how it does its job. Sometimes the story is simple. The twentieth-century US Supreme Court Justice Louis Brandeis once wrote, in juridical context, that 'Sunlight is said to be the best of disinfectants'. To students of infectious disease his statement is more than metaphor. Many pathogens die when exposed to ultraviolet because it damages their DNA. Its success first became obvious when used against the facial form of tuberculosis and its power is also manifest in the widespread use of ultraviolet light to sterilise drinking water. The spread of drug resistance has led to new interest in the possibility of its broader role in the fight against contagion.

Moderate doses of ultraviolet may help wounds to heal, particularly after surgery. It sterilises the incision and causes skin cells to make chemical messengers that speed up repair. Another scheme uses brief flashes of intense radiation that allows our own DNA to repair itself but destroys that of bacteria and viruses. Yet another uses fluorescent dyes taken up by bacteria. A short burst of light can then kill them. The process also helps against skin ulcers caused by parasites.

The feeblest form of ultraviolet, UVC, does not make it through the atmosphere, but can be generated with a laser. Its waves are absorbed by DNA and it kills many bacteria. Even blue light alone can do a lot to help, and is used in the treatment of acne and psoriasis. Light therapy of this kind has the further advantage that it has an effect within an hour or so, while medicines may take days to do their job.

In the days before penicillin, doctors sometimes took blood from a patient with an infectious disease, irradiated it with ultraviolet and returned it to its owner. The initial work was done in the 1920s with dogs ill with blood poisoning. It was at first assumed that the radiation simply killed off the agent responsible, but it was then found that to treat just one part in twenty of the animal's blood had the same effect. The technique also worked well on human patients, some almost at the point of death. Quite how it does the job was, and still is, not certain. Then antibiotics emerged, and the technique became the cure that time forgot. Now may be the moment to recall it.

Viruses, too, are susceptible to the wonders of sunshine. Epidemics of colds and flu are more common in winter than in summer. People do crowd together then, but the pattern emerges in part because viruses survive better in the gloomy days of December. A shortage of daylight also reduces the effectiveness of the immune system as vitamin levels drop. Those who take a regular vitamin supplement have fewer cases than others, and if their habit became universal, three million Britons would avoid such annoyances each year. The potential of light as a defence against human disease, once almost forgotten, is back on the front line.

The ability of ultraviolet to destroy pathogens exposed directly to its rays is easy to understand. Its influence on infections within the body is more subtle. Some comes from the ability of the Vitamin D to switch on the body's natural defences. The vitamin pushes up the activity of a gene that codes for a small protein called cathelicidin. This kills off bacilli in the bloodstream and elsewhere – the agent of tuberculosis included – because it can punch its way through their cell membranes. The vitamin can slow or reverse the symptoms of illness, but a shortage causes the antimicrobial gene to switch off and the problems to continue. Tuberculosis, like rickets but for quite a different reason, is hence in part a vitamin-deficiency disease.

A day under bright blue skies does more than fight infection. The skin is an extended endocrine organ that makes a variety of chemical messengers when exposed to solar rays. They include precursors of steroid hormones such as oestrogen and testosterone, together with substances such as serotonin that alter its bearers' frame of mind. Oxytocin, the so-called 'love hormone' associated with pair-bonding, endocannabinoids, responsible for the 'exercise high' reported by athletes, and endorphins, a family of chemicals that can reduce pain and help maintain a calm disposition, are also produced. George Harrison, Robert Louis Stevenson, my undergraduate self and millions of today's holidaymakers have, without realising as much, generated a complex web of internal messengers to cheer themselves up whenever they sought its rays.

As is true for all mood-altering drugs, some people indulge in too much of a good thing and fall prey to an almost pathological need for sunlight. Around one in five young white American women uses a sunbed at least once a year, and quite a few lie on them almost every day. Heavy users say that they feel driven to submit to temptation even if they feel guilty about the habit, and that they have tried, and failed, to cut down. They are very sensitive to the presence of the magic beams, and can tell at once the difference between a sunbed doctored to emit no ultraviolet and one in full working order. They lose that ability when given a drug that blocks receptors for mood-altering hormones. All those attributes are shared with those addicted to other narcotics.

Ultraviolet also generates the nerve transmitter dopamine. One of its many roles is to produce a positive response to rewards of various kinds and as many addictions – that to sunbeds perhaps included – become entangled with it. A shortage is also associated with some cases of attention deficit hyperactivity disorder, ADHD, which may be why this condition is less common in sunny places. It also alters patterns of growth of the eye in a manner that depends on light intensity. The developed world has seen a considerable jump in the incidence of myopia – near-sightedness – over the past three decades and in Japan the majority of children have the problem. In parts of China, where it is also rife, infants are now forced to abandon their schoolbooks and screens and to go outside for an extra hour a day in the hope of reversing the trend.

Perhaps the most unexpected compound to be born in sunlight is a simple chemical now known to have complex effects. Nitric oxide – one nitrogen linked to an oxygen – acts as an internal messenger that works fast and over short distances. The compound relaxes blood vessels and lowers blood pressure. Drugs such as amyl nitrite, which fight circulation problems (and are used in 'poppers' to generate euphoria), produce the stuff, as does the sexual stimulant sildenafil (otherwise known as Viagra). Nitric oxide is active in the circulation, in the immune system, in the fight against infection, in the control of blood sugar, and in the defences against obesity and diabetes.

The skin is rich in compounds that contain both nitrogen and oxygen, with a store ten times greater than that held in the blood. In sunlight large amounts of nitric oxide are generated. This relaxes the walls of arteries, allows blood to flow to the skin and drops blood pressure, which is at lower levels in summer than in winter, and in sunny places compared with cloudy, so much so that some people with high blood pressure benefit from an hour-long session in the open air. The treatment is not guaranteed, but is safer than drugs and is at least enjoyable.

All this means that phototherapy – treatment of diseases with light – has begun to return to fashion. Some modern machines use dozens of different bulbs. Those at the blue end kill bacteria and stimulate the production of nitric oxide, which widens blood vessels and may help to relieve pain, while the longer waves promote the formation of molecules that help heal wounds and ulcers, and ultraviolet kills off infections. The devices have been used with some success for the treatment of ulcers and bedsores.

In spite of such advances, not everything is sunny in the world of solar medicine. The field is filled with disagreement, some of which verges on the rabid. Too often, optimism has raised its hoary head, only to be countered by almost immovable pessimism. A remarkable variety of conditions has been blamed on a lack of ultraviolet. They include – among others, and in alphabetical order – acne, allergies, Alzheimer's disease, asthma, attention deficit hyperactivity disorder, autism, back pain, bronchitis, cancers of a dozen kinds, chronic fatigue syndrome, constipation, dementia, dental caries, depression, dermatitis, diabetes, diarrhoea, eczema, endometriosis, falls and broken bones, gout, gum disease, high blood pressure, inflammatory bowel syndrome, jaundice, kidney stones, low birth weight, lupus erythematosus (an inflammatory condition due to an attack by a patient's own immune system), mononucleosis (the viral 'kissing disease' of teenagers), multiple sclerosis, obesity, obsessive compulsive disorder, osteoporosis, Parkinson's disease, pneumonia, pregnancy failure, psoriasis, rheumatoid arthritis, schizophrenia and sepsis. Today's emergence of sunlight as a universal remedy often has, the doubters suggest, as little support as the ancient Egyptian claim that it prevents baldness.

Some apparent fits of poor health with lack of sunlight are without doubt spurious. Low vitamin levels and knee surgery go together, but that is because patients on crutches find it hard to go for walks in the open air. The same is true for the many illnesses that involve a forced stay in bed, or long periods indoors.

One common – and perhaps sometimes justified – complaint by sceptics is that surveys often find no difference in the proportion of people who fall below the recommended levels of the vitamin and their risk of heart disease, multiple sclerosis or diabetes. That may be true, but much of the work has been done in places in which most individuals have reasonable levels of the compound as measured on the official scale used to assess childhood bone health, but may not have enough to influence diseases of adult life. In addition, a childhood episode of deficiency, later reversed, might still cause ill-health in adults. Undue caution also plays a part, for in the heroic days of vitamin treatment large amounts were given, but many of today's treatments use far less on – probably unnecessary – safety grounds.

Attempts to improve the prospects of patients with a variety of conditions with doses of the magic compound have also given equivocal results. A recent survey of thousands of adults who take a supplement shows no effect on bone health – although, once again, most of them had reasonable levels even without an extra pill, and the positive effects, if any, on other diseases were not examined. A rigorous test needs a randomised trial, in which groups are given pills that either do, or do not, contain the active component, with nobody told until the end who received what.

Whatever the disagreements within the medical profession, public interest in the substance has rocketed. The number of blood tests among older Americans has gone up by eighty times since the turn of the present century, and around one in five adults there take a dose every day, sometimes a dose forty times that generally recommended.

Until the 1950s, most European authorities, those in Britain included, supplemented food with the stuff. They added it to the yeast used to make bread, and even, in Britain, to beer and custard, or fed hens or cattle with it to push up the levels on the plate. It was then banned because of what we now know to be an unjustified fear of its supposed toxicity.

Soon that concern was dismissed and, in many places, the supplements came back. The European Union authorities now suggest that adults should take in fifteen micrograms of Vitamin D a day, which for many people, especially in winter, would call for an artificial boost. Until not long ago, Britain had the dubious distinction of being the only country not to make an official recommendation about whether to take one at all. In 2016, for the first time, Public Health England suggested that everyone over four should take a supplement of ten micrograms every twenty-four hours. This may not go far enough, for the infants of women who depend on breast-feeding alone almost never have enough. Either their mothers should be given more supplements, or – a simpler solution – the baby should from time to time be fed fortified formula milk. Most mothers do not follow that advice, and some doctors feel that it should, like vaccination, be automatically provided by the health service just after a baby is born.

In the United States and Canada, a scheme to add Vitamin D to milk, margarine, cereals and baby formula, fruit juices and yoghurt began soon after the war ended, and such foods are now an important source, for the poor most of all, with new plans to double the amount added to milk. Plenty of North Americans still face deficiency, but without the artificial extras many more would have that problem. In Europe, Finland has also had great success, with almost the whole adult population at satisfactory levels. India, too, with nine-tenths of the population deficient in some places, is considering such a programme for the government-subsidised rice, oil and milk eaten by children, hospital patients and the poorest.

Germany, in contrast, still holds to a seventy-five-year-old food-purity law that limits addition to any food except margarine. Britain has also lagged behind, for just margarine and a few cereals are fortified as a matter of routine. After the fall of the Attlee government, a hatred of state intervention loomed large. There was particular opposition to adding the vitamin to milk, which to some politicians seemed to have an almost mythical purity. A few companies do now add it to yoghurt, powdered milk, bread and breakfast cereal, but the North American approach has been more effective. To add even small amounts to flour would, in this nation of bread eaters, halve the numbers who take in less than they should. Mothers are advised by some doctors to offer supplements to their infants, but simple advice is not enough, for the success rate is much lower than in the many European countries that routinely add the compound to foodstuffs. In modern Britain, money talks loudest, and a fortification programme would, according to some estimates, cost only a tenth of the amount saved by the National Health Service by having to mend fewer broken bones and to deal with other effects of shortage.

As the parliamentary question and answer quoted at the head of this chapter suggests, in spite of the recent rise in the frequency of the disease our rulers still maintain the lordly detachment of their predecessors. Not much has happened, it seems, to change their party's mind since the post-war debates about the need for intervention in public health.

Whatever the joys of supplements, the most convenient source of Vitamin D is, as it always has been, sunshine, and one immediate cause of the return of the condition has been today's tendency to avoid its rays. In some places, people have no choice.

This book began with a disparaging account of the Scottish climate. Plenty of Scots complain about it, but most do not realise how exceptional their weather in fact is. Fewer than one in fifty people worldwide lives as far north as they do, and just one in two hundred lives nearer the pole than do the inhabitants of Shetland. Edinburgh and Glasgow are close to the northern limit of all cities with a population of more than half a million. Their position on the western edge of Europe gives them the added benefit of the Gulf Stream and its clouds, and means that those who live there see fewer sunbeams than do those of almost any other large city. Stockholm, for example, is two hundred kilometres north of Edinburgh (and with almost a million people is the closest metropolis to the North Pole), but gets sixty per cent more ultraviolet. Glasgow does even worse, for it receives no more sunlight than does the far north of Sweden, a thousand kilometres further towards the Arctic than the country's capital. The 'Vitamin D winter' – the period in which nobody makes enough of the stuff – lasts a month longer on the banks of the Clyde than it does on those of the Thames. To add to the problem, teenage Scots sit at the bottom of the global league table of how long they spend outside each day.

In Scotland, deprivation began to rise in the 1980s when heavy industry began to collapse. Even so, its health issues do not come just from hardship, for Liverpool and Manchester have the same levels of that as does Glasgow, while there are about a third more premature deaths in the Scottish city than in either of the others. The excess applies to both sexes, to rich and to poor, and persists even when corrected for Scotland's higher levels of obesity and tobacco use.

Almost all the 'diseases of darkness' – rickets, drunkenness and depression included – are more prevalent north of the border than they are to the south (tuberculosis is the exception, with the rate in cosmopolitan London fifteen times higher than in Glasgow). Male life expectancy in Scotland when I arrived there in 1962 was sixty-six, compared with the English figure of sixty-eight. In both places it has now risen by more than a decade, but Scottish men still live on average for two years less than do Englishmen, and the gap is widening. In the European context, England and Wales have themselves not done particularly well, for after a century of steady progress there has been almost no improvement in life expectancy since 2010. Then, they were at the head of the table of annual improvement; now, they are almost at the bottom. Scotland has done even worse, for in recent years the average age of death has not risen but fallen. Just Bulgaria, the Czech Republic, Estonia, Hungary, Latvia, Poland, Romania and Slovakia have a lower life expectancy, and in all of them the trend is upwards. As a result, most countries in Europe may soon overtake it.

To see shocking disparities in health there is no need to leave Scotland's largest city. When mapped as one-kilometre squares, the ten most deprived areas in Britain are all in Glasgow. Men in the inner-city district of Calton live on average for twenty-eight years less than do their fellows in the leafy suburb of Lenzie, twelve kilometres away. Across the city as a whole, the physical and mental condition of the most, and the least, deprived sixths of the community shows a burden of disability in the former that begins in childhood and continues through life. Men in the poorest cohort have an average age at death of just under seventy, while those in the richest last for a decade longer.

Parts of Scottish society have long had a drinking culture, and Boswell's *Edinburgh Journals* give a hair-raising account of the extent to which he, a successful lawyer, was completely inebriated several times a week. The difference in alcohol-related deaths north and south of the border reached a peak early in the present century, at twenty-two per cent. It has declined since then, but the improvement stopped in 2013 and is on the rise once more. In 2018 a minimum price for alcohol was introduced, but it is still too soon to see if this will have an effect. In addition the country has a death rate from drug overdose twice that of England and higher than anywhere else in Europe.

In addition to all those challenges, low vitamin levels also play a part in its relatively poor health. In Glasgow, the most affluent sixth of the population almost all have a level of Vitamin D at or above the levels recommended as adequate, with many individuals well above it and just a small minority below. For the poorest, the average is lower than the threshold, with many well below the safe level and some with almost undetectable amounts.

Even children born in Scotland who move south in their teens die younger than their fellows, perhaps because a childhood shortage has lifelong effects. Rickets is still commoner there than in equivalent communities in England, and bone fractures in children happen at one and a half times the southern rate. The incidence of multiple sclerosis in England is around a hundred and sixty in every hundred thousand, and in the mainland of Scotland about two hundred and ten. That figure rises to three hundred in Orkney and four hundred in Shetland. Deaths from heart disease north of the border are a quarter higher than in England. Cancer mortality, too, is somewhat greater. Suicides are also a concern, with a mortality three-quarters higher than in the south, perhaps because the dim days of an Edinburgh January are enough to depress even the most robust.

The authorities have begun, rather late in the day, to deal with the effects of a shortage of sunshine. In 2016 the Scottish government recommended, in parallel with the health authorities in England, that all residents over a year old should take ten micrograms of Vitamin D every day. It has gone further than its southern neighbour, for it offers free supplies to all pregnant women, and to those on benefits with children under four. In England, in contrast, free supplements are restricted to pregnant women on benefits who are more than ten weeks pregnant or have other children under four. In both places, the contrast between the attitudes of our rulers to their nation's physical and mental state and that of their predecessors in the poverty-ridden post-war era is stark.

The centre founded to deal with the vitamin deficiencies of those days still stands in Finsbury, now part of the London Borough of Islington. A district that was then a byword for hardship has been transformed. It has become the new

Chelsea, awash with the cash of those who work in the City (and – I have to admit it – I lived nearby for two decades). Its architect, Berthold Lubetkin, also designed several blocks of workers' flats in the borough. They have become monuments of modernist architecture. His Health Centre, in contrast, is rather run-down. Trees have grown up around it to muddy its elegant lines, and it now hosts no more than a doctor's surgery, a smoking clinic, a Centre for Sexual and Reproductive Health, and a centre for childhood stuttering. The 'electrical treatment clinic' that used ultraviolet to ward off bone disease has gone, as have the tuberculosis consulting rooms and the murals that enjoined visitors to 'live out of doors as much as you can' and enjoy 'fresh air night and day' (not easy in the days when the skies were black with smoke). A scheme a decade and a half ago to sell the place for conversion into luxury apartments was scrapped after public protest, and the Finsbury building now stands as a rebuke to, rather than a celebration of, later governments' failed attempts to build on the post-war triumph in public health. As Lubetkin himself said, 'These buildings cry out for a new world that has never come into being'.

As the war ended, his vision of a new society built on socialist principles began to fade. Soon after the Iron Curtain split Europe, his last council-block scheme, under construction on the bombed site of Lenin's old lodgings, which he had planned to call Lenin Court, suffered a two-letter change of identity and became Bevin Court instead (Ernest Bevin, then Foreign Secretary, was intensely anti-Soviet). Infuriated, the architect buried his monument to the Russian politician in its foundations rather than, as he had hoped, placing it on display in its entrance hall. Many of the flats within Bevin Court are now in private ownership (and the whole block, with its hundred and fifty units, was built for the present price of a one-bedroom apartment).

Five years after Lenin (or Bevin) Court opened, Lubetkin was asked to help with the development of Stevenage New Town. He declined and in an embittered letter wrote that:

> The golden age of modern architecture is over and done with: its current trivialities have become dreary and ineffective . . . I have decided to stick to my original decision to step out of this 'profession' and no posters calling for a new Britain on the background of an anaemic photograph of our Finsbury Health Centre will persuade me otherwise.

He was involved for a short time in a scheme for a New Town in County Durham, but that too failed. Soon, his career as an architect faded and he became a pig farmer. Lubetkin's dream was, perhaps, in part vindicated when in 1982, at the age of eighty, he received the Gold Medal from the Royal Institute of British Architecture.

I, on the other hand, have after completing this chapter begun to take Vitamin D supplements. I buy them over the counter rather than being a drain on the state. The political descendants of Winston Churchill would no doubt approve.

CHAPTER 6

THE CHAIN OF THE HOURS

Yet of this change so frequent, so great, so general, and so necessary, no searcher has yet found either the efficient or final cause; or can tell by what power the mind and body is thus chained down in irresistible stupefaction.

Dr Johnson on Sleep (*The Idler*, 1758)

Somewhere in the world, the sun is always shining, but to its citizens, wherever they live, the star's most obvious property is that it disappears for half the time and their lives shut down in sympathy. Why do we fall into a coma for several hours each day? Until not long ago we had no idea, but now we know.

Marcel Proust is literature's greatest prophet of sleep, and of memory. He is the author of the most quoted (or most hackneyed) sentence in French literature:

Longtemps, je me suis couché de bonne heure. Parfois, à peine ma bougie éteinte, mes yeux se fermaient si vite que je n'avais pas

> *le temps de me dire: 'Je m'endors.'* – For a long time I used to go to bed early. Sometimes, when I put out my candle, my eyes closed so quickly that I did not have time to say to myself: 'I'm falling asleep.'

Such are the opening lines of his exploration of the nature of lost time, *À la Recherche du Temps Perdu – In Search of Lost Time* – published in seven volumes (three of them posthumous) from 1913. The work is held together by recollection; by the sudden and unexpected return of moments that bring the past and the present together (and as far as I recall, scented cakes of some description are also involved).

In its first paragraph the narrator – who shares a Christian name (and perhaps more) with the author – looks back at his childhood ability to plunge into the land of dreams as soon as the light went out, only to wake in a trance-like state, convinced that he is a character in the book he had taken to bed. He was not alone in that experience. Paul McCartney claims to have composed 'Yesterday' while asleep; as he said: 'I just woke up one morning and it was in my head' (his imagined first line, on the other hand, began with 'Scrambled eggs . . . '). George II, too, while in bed once saw a vision of his deceased wife, Caroline, and was so disturbed by the experience that he summoned a coach in the dead of night to take him to her tomb in Westminster Abbey.

In a more pedestrian experience, the Austrian biologist Otto Loewi noted in the 1920s that a frog heart in a nutrient liquid slowed down when its nerve was stimulated. This was assumed to be due to a simple electrical message, but Loewi dreamt that he had put a second heart from which the nerves had been removed into the liquid as well. He did the experiment as soon as he awoke, and at once the second heart too slowed, as proof that information was also transmitted by a chemical. That result is now central to brain science, which has discovered dozens of such nerve transmitters.

The unstable frontier between dreams and reality is a key to *In Search of Lost Time*. What was in its author's day an undiscovered country has begun to yield to science. What lies behind the regular rhythms of what Proust called 'the chain of the hours, the sequence of the years, the order of the heavenly host'?

Slumber once seemed no more than an economical way to spend the hours of darkness; as Dr Johnson put it: 'Once in four and twenty hours, the gay and the gloomy, the witty and the dull, the clamorous and the silent, the busy and the idle, are all overpowered by the gentle tyrant, and all lie down in the equality of Sleep.' An 1834 medical text *The Philosophy of Sleep* regarded the interlude in entirely negative terms: 'Sleep is the intermediate state between wakefulness and death: wakefulness being regarded as the active state of all the animal and intellectual functions, and death as that of their total suspension.' That view has long been rejected by biologists. Even so, the regular plunge into oblivion was until not long ago unique among the body's functions in how little was understood about it.

A pattern of rest and activity driven by the sun emerged with the first cells. As life unfolded, that process was hijacked by a host of other functions. In evolution, that often happens.

The first lungs developed from cell membranes able to extract oxygen from water (and the lung surface must still be kept wet to allow it to work), but now they help to keep their owners cool, control the chemistry of the blood and even provide the power of speech. Sleep has been invaded by an even greater range of functions, but what came first, and why, how or even whether any among them need a loss of consciousness was until not long ago quite obscure.

Johnson's gentle tyrant is just one of a series of internal cycles that pervade body and mind. Such circadian rhythms, as they are called, provide silent reminders about what should come next, for we feel tired, hungry or lubricious at about the same time each day, which is at least convenient (one of my own sentinels prompts me when to pour the first glass of wine). Such behaviours are regulated by dawn and dusk, and when – as nowadays they so often do – men and women ignore their instructions the effects may be dire.

Rhythms pervade the body, for at least half of our genes show changes in activity as the hours wear on. They are more than mere alarm clocks. The time spent in bed repairs the damage done each day to the body's frame, alters hormone levels, determines how much fat is laid down and primes the immune system. It is also an unexpected key to a talent that we all share: the ability to store memories for days, months, or years. Proust's ability to search for, and to recover, the past had, although he never knew it, a direct tie with the hours he spent between the sheets.

French authors slumber, but so do fruit-flies and octopi. Even amoebae snooze, for at times they draw in their tentacles and form a ball. If cruelly prodded and woken up they stay active for a while, but soon, like a tired baby, fall back into inertia. Bacteria can scarcely be said to sleep, but some do have rhythms of activity. The marine ancestors of today's blue-green algae – the cyanobacteria – were the first to soak up sunlight and to generate free oxygen long ago. Their descendants still have the oscillators that primed those ancient cells to wake up as dawn approached.

The world of repose has many quirks. Giraffes shut their eyes in short bursts for just two hours a day, while some migratory birds nod off on the wing for seconds at a time. In contrast, a certain North American bat needs twenty hours (as does, for that matter, the two-toed sloth). Dolphins send half their brain into inactivity while the other half stays alert until it becomes time to take a rest, so that one eye is always on the lookout. Several birds do the same – but mallard ducks go further, for they line up when the sun goes down. The two on each end of the row keep the outermost eye open, while those in the centre can afford to close both.

Homo sapiens is the wakeful primate. Modern life is not to blame, for hunter-gatherers sleep about as much as most Europeans, although they retire around three hours after sunset and rise before dawn. A chimpanzee makes a new bed of branches each night in a tree. It closes its eyes soon after dark, to wake at sunrise, eleven hours later (the owl monkey needs six hours more). Human restiveness may have begun when our ancestors came down from the trees. A new life on the ground demanded more of a lookout for enemies, both human and animal. As numbers grew, extra time was needed to chat with the neighbours (or to pick off their fleas), and hunters may have had to work for longer each day to find food for larger and larger groups. Our slumbers may be short, but to make up for that they are deeper than those of our relatives.

A third of every life is spent in bed. A new-born baby stays insensible for eighteen hours in each twenty-four, with brief blocks of wakefulness brought on by hunger. Five-year-olds need eleven hours, ten-year-olds ten hours, and those in their mid-teens should get nine. The amount called for continues to reduce with age, but a person of average life expectancy (now around eighty) will still have spent a quarter of a century with eyes tight shut before they close for ever.

Just one person in three in the developed world hits those targets, and the World Health Organization has declared insomnia to be an epidemic. Even so, at seven and a half hours a night the British are the drowsiest people in Europe apart from the Finns and the Dutch. The Chinese used to do even better, for many of them took a three-hour lunch break with a nap included, which gave them nine hours of rest a day, but now the authorities have cut out the lunch-time snooze.

Everywhere, the rich sleep better than the poor. Exceptions do exist: Mrs Thatcher slept for four hours a night ('Sleeping is for wimps!'), while Donald Trump claims that he needs just three. Our first female Prime Minister showed signs of dementia in old age, as does, according to a book by one of his dismissed aides, the present leader of the United States. At the other end of the political scale, President Calvin

Coolidge ('Silent Cal') slumbered for eleven hours in every twenty-four.

A single block of torpor seems normal to Western eyes, but until not long ago an afternoon siesta was the rule in Spain and other places. Before the invention of electric light English diaries, too, sometimes mention the 'first sleep' and the 'second sleep'. Most people snoozed for several hours after an early dinner and woke for an hour or so at midnight, in the 'night watches'. They were exhorted to use the time to consider their mortality and the goodness of God, because, in the words of the seventeenth-century parliamentarian and mystic William Killigrew: 'regenerate man finds no time so fit to raise his soul to Heaven, as when he awakes at mid-night' (a few souls may have done so, but many no doubt cleaned the kitchen instead). African subsistence farmers still follow that routine, so that what seems normal today may be a departure from the natural pattern. Among the Hadza people of northern Tanzania, many of whom are hunter-gatherers, individual habits vary so much that in a typical group of thirty or so there are, perhaps as a defence strategy, almost no moments in which all members are deep in slumber.

Brief naps, too, can be restorative. Winston Churchill took several throughout the day and night, while Ronald Reagan was also fond of them and once asked a journalist to wake him up at a set time 'even if it is in the middle of a cabinet meeting'. In most of Europe the siesta has been abandoned. When I first went to Greece fifty years ago almost all shops were closed for four hours in the afternoon as their owners snored behind the shutters. That practice has disappeared and the once widespread claim that those who snoozed had lower levels of heart disease has failed to stand up.

Life has always bowed to the rule of the sun. It responds to light and dark, to the intensity of its rays, and to its wavelength. In mice, a change from red to blue and back again sets the clock almost as well as does the rhythm of night and day. For men and women light with a blue tinge (typical of dawn) interferes more with the ability to drift off than does the reddish hue of sunset. The body also responds to regular shifts in temperature, in noise and in mealtimes, while for some reason low oxygen levels, as on a long flight, activate some of the genes involved in the body's clock.

Why do we sleep? Some theories are fanciful, with dreams seen as messages from the gods (or the psyche), while others are quasi-scientific, but wrong. For much of history, the nocturnal oblivion was seen as a digestif, an aid to digestion. As night drew in, the heat of the body was directed towards the stomach. There, as in a saucepan, it began to soften the food. Savoury fumes rose to the brain, itself cold after the efforts of the day. Soon the vapours cooled down and dispersed throughout the body. As they did, they reduced the choleric spirit that built up in daylight hours and lulled its owner into insensibility; as Sir Thomas Elyot's lines of 1539 put it: 'Naturall heate, which is occupied about the matter, wherof precedeth nouryshment, is comforted in the places of digestion, and so digestion is made better, or more perfite by slepe, the body fatter, the mynde more quiet and clere, the humours temperate.' Some of that does hint at the truth, but digestion in fact slows at night to get its breath back before it has to cope with the challenge of breakfast.

The centre of attention then switched from stomach to skull. A century after Elyot's day, Thomas Willis, one of the founders of the Royal Society, injected a dye into the arteries of the brain and noted that their structure would allow a constant exchange of blood with the body. René Descartes came up with a hydraulic model of the process, in which the pineal gland at its base (in his view the site of the soul) caused the liquid-filled ventricles of the grey matter to collapse at bedtime and the brain to shut down. That, too, was plausible but wrong.

Science has at last begun to uncover the truth, which is more complex and more unexpected than once seemed reasonable. Its progress was recognised by the award of the 2017 Nobel Prize in Medicine to three of those involved in the research.

The body's clocks take most of their information from receptors in the retina. The familiar rods and cones assess the intensity, and the colour, of light with the help of pigments called opsins, which turn photons into electrical signals. Another member of that family of genes, melanopsin, was discovered two decades ago in a study of the dark blotches on the skin of the African clawed frog, which change size in response to sun and shade. That seemed at first little more than a curiosity, but the frog's eye receptor was then found in our own eyes. There it assesses the intensity of light, most of all that at the blue end of the spectrum, and helps to set the internal clock. The rod and cone opsins also play a part, for mice must be mutated to lack all three types before their timer loses all contact with the outside world.

The move into and out of alertness is in part a matter of chemistry. The level of a compound called adenosine increases as dawn moves towards dusk until it reaches a critical level. It then binds to nerve cells that induce slumber, and they start to shut the system down (the molecule is secreted after sex to much the same effect). A large espresso three hours before bedtime delays the ability to nod off by about an hour. In a well-lit room that delay almost doubles, so that an evening coffee-drinker in London in effect moves to Reykjavik in a drug-induced jet-lag.

As perception fades, body temperature drops and levels of hormones begin to change. One among them, melatonin, is secreted by the pineal gland and attaches itself to receptors scattered throughout the brain. The substance is produced as the light grows dim and sends a signal that bedtime is close. Depending on the length of day, and the habits of its owner, its concentration rises fast after about nine in the evening, and reaches a peak in the middle of the night, leaving the stage about twelve hours later. Because winter nights are longer than those in June, melatonin is also a cue to season.

Not everyone obeys its instructions. A baby's internal timer is liable to turn on and off at what seem like random intervals. It begins to settle down after the infant reaches more than about three months old. Many adults, too, choose to ignore its advice, but if they overdo their rebellious behaviour both mind and body will suffer.

Once in the arms of Morpheus, tics and signs obvious to any bed mate hint that the experience involves rather more than oblivion. Sleepers mutter, or come out with complete sentences, grind their teeth, swap from side to side, and snore (men more than women). For a time they may lie tense and almost rigid (penis, if available, included), while in other periods they slump into paralysis. The two patterns repeat themselves in a cycle, several times a night. Then at last the subject begins to stir, to mumble, to wake, and to stagger out to face the world.

Such activities are symptoms of inner events. Growth hormone is secreted during sleep, as is its relative, prolactin (initially named for its effect on milk secretion). Each pushes up the rate of cell division. The hormones that lead to satiety reduce their activity, which sharpens the appetite for breakfast, and others suppress the desire to urinate, so that most people can manage eight hours without a trip to the bathroom. Levels of the steroid hormone cortisol drop to a low point at around midnight and peak just before the return to wakefulness, to alert the body to the excitements to come. All this means that the internal economy of a person deep in slumber is quite different from that of the same individual at midday.

The real drama takes place within the brain. In daylight its cells generate a choppy electrical sea as they cope with the world around them. As its owner enters the world of repose, the weather begins to calm down. Once comatose, the grey matter generates billows of slow activity, with occasional local squalls.

Then comes a sudden change. A hurricane sweeps through the skull. Its paradoxical combination of a flaccid body accompanied by mental uproar is known as 'Rapid Eye Movement' (or REM) sleep. The subject might seem insensible but his or her brain is talking, or even shouting, to itself, so much so that some parts are more active than they are when awake. Then the storm abates. The REM episode is replaced by non-REM, muscles regain their tone and the brain is once more satisfied with slow surges accompanied by faster movements within a narrow range. A typical night involves four to six such cycles. In the first hours in bed, non-REM slumber dominates, but as the night wears on, REM begins to take over. It was once assumed that dreams took place only in REM sleep, but that was not correct. A small 'hot zone' in the cortex becomes more active as a subject dreams, and does so in both forms of slumber.

In some ways, the two patterns of repose are as distinct as are sleeping and waking themselves. Each is found in most mammals and birds, but full-fledged REM sleep is rare in other animals, proof perhaps that it emerged with the increased intelligence of ourselves and our feathered friends. Among primates, humans have the most of all. Perhaps this began when we left the trees for the ground, for to have a flaccid body while comatose in a nest high above the surface is a dangerous luxury. Once on the ground, or in a four-poster, it can be indulged in with no such risk.

Sleep does not, as is often assumed, mean complete separation from reality. When between the sheets, the brain responds more readily to the sound of its owner's name than to someone else's, or to an angry rather than a placid voice, as it does when its proprietor is awake. Dreams, too, blur the boundaries between reality and imagination. They have long been a source for mystical and quasi-scientific attempts to understand the mind. Theories have come and gone but now the truth has begun to emerge.

Nocturnal visions were a useful source for the Church's attempts to summon up spectres that it alone had the power to banish. Rib muscles lose their tension in REM sleep, and it becomes harder to breathe, perhaps because a creature – an evil spirit – has landed on the chest. The signs of sexual excitement that sometimes manifest themselves add weight to the claim that an incubus or succubus has appeared. Dr Johnson defined 'nightmare' as 'A morbid oppression in the night resembling the pressure of weight upon the breast' as a hint that he had experienced the sensation (a dream was a mere 'phantasm of sleep'). So fiendish are such evil spirits that they render the body immobile to prevent the afflicted from protecting themselves with sacred gestures. Holy relics around the bed can, the experts say, keep them at bay.

Thomas de Quincey, author of the 1821 *Confessions of an English Opium Eater*, tells of his own drug-fuelled visions in which he was 'kissed, with cancerous kisses, by crocodiles' (a sensation, I am forced to admit, that I have never shared), but gives a literary rather than a theological insight into their value. For him, dreams were 'the one great tube through which man communicates with the shadowy . . . and throws dark reflections from eternities below all life upon the mirror of the sleeping mind'. Another unlikely – and as an adult almost as eccentric – proponent of the recounted dream was the young Princess Margaret, who, her nurse Crawfie tells us, would hold her sister Elizabeth enthralled with accounts of talking cats and the like that sometimes went into several instalments.

De Quincey was an avid consumer of laudanum, opium dissolved in wine or brandy, and wrote of his first experience of it that: 'portable ecstacies might be corked up in a pint bottle: and peace of mind could be sent down in gallons by the mail coach'. A derivative of that drug called oxycodone was developed in Germany at the time of the First World War and was regularly used by Adolf Hitler during the next great conflict. It is now at the centre of an American drug epidemic which has killed tens of thousands.

As with Proust – who wrote much of his work while addled by his own daily sleeping draught – for Hitler, and for de Quincey, the boundaries of night and day, of slumber and alertness, became blurred. Seated in what had been Wordsworth's cottage in the Lake District, de Quincey did not come into his own until after midnight, when he entertained (or infuriated) his guests with long tales of his drug-induced dreams. Oscar Wilde once remarked that the most frightening sentence in the English language was: 'I had an interesting dream last night' and de Quincey's listeners may have shared that opinion. The habit continued throughout his life, and one of his acquaintances noted that in the author's troubled later years in Edinburgh 'the first difficulty was to get him to visit you; the second was to reconcile him to leaving'. By coincidence, his last days were spent in a house a few doors away from that earlier occupied by Darwin. It survives, but is free of any form of memorial.

De Quincey also wrote – correctly – that: 'The machinery for dreaming planted in the human brain was not planted for nothing'. His own nocturnal fantasies were of crocodiles, although without doubt he had never met one (I myself do admit to some episodes that involve snails). Galen, the Greek physician of the second century, believed that their contents reflected the four humours of the body, Earth, Air, Fire and Water, and would be useful in diagnosis: 'Someone dreaming a conflagration is troubled by yellow bile, but if he dreams of smoke, or mist, or deep darkness, by black bile. Rainstorm indicates that cold moisture abounds; snow, ice, and hail, cold phlegm.'

Most such excursions are more banal. Men dream more of male strangers and of the open air than do their female partners, who prefer visions of friends of their own sex and of life at home, while babies hallucinate for half their hours of slumber, although we have no idea about what. For adults that figure drops by half, and continues to fall, so that the sleep of old age brings forth fewer monsters than does that of youth.

Nightmares can be disturbing, to children most of all (and I still remember a gruesome one at the age of ten or so when I had for some reason been immersed in a book on cannibalism). The idea that they represent imbalances in the body's humours, or messages from the spirit world, was succeeded in the nineteenth century by Kant's claim that the world of a dreamer resembles the waking life of the mentally ill. The idea was taken further by Freud, who saw such episodes as hints of hidden emotions buried in the unconscious mind that, if identified, would, given a long (and expensive) period of analysis, say a lot about our state of mind when awake. That notion was described by the UCL immunologist Peter Medawar as 'the most stupendous intellectual confidence trick of the twentieth century'.

Freud – himself addicted to cocaine – in *The Interpretation of Dreams* was nevertheless the first to draw a parallel between the visions seen in dreams and in psychotic episodes and those summoned up by the abuse of various drugs. In such situations it becomes impossible to separate the real from the imagined, with images of devils or monsters, of impossible tasks, and of the need to carry out violent or dangerous acts to achieve them. Dreamers and psychotics each suffer wild fluctuations in mood, and a loss of the senses of self, of time and of place. For both, when the episode is over there remains almost no recall that it ever took place.

The shifts in sight, sound and emotion as a dream goes on show that the brain is in control of all its senses except one. Only rationality has been suspended. Brain scans of those who experience 'lucid dreaming' – in which the dreamer is in some senses awake but is aware that his sensations are in part unreal – show that the sections involved are also active in patients who suffer a psychotic event, although quite why that is we are not sure.

Dreams do at least one useful job. The brain's version of the 'fight or flight' hormone adrenalin works hard to promote vigilance and alertness. When faced with a sudden stress its levels rise and increase the level of anxiety. In the REM phase its supplies are shut off. Perhaps this allows memories of fear and rage to be scrubbed clean and to be recalled in sanitised form. Dr Johnson noted the effect when he wrote that 'Memory is the purveyor of reason'.

When I was attacked and robbed in Camden Town many years ago (an event about which I can summon up a vivid recollection) I hit one of my two assailants, breaking a finger, and chased after both of them screaming as hard as I could. Perhaps fortunately, they got away. When I got back from the Accident and Emergency Department at University College Hospital I was still quite shaken. The next day, as I was about to go to work, to my own amazement I picked up a kitchen knife and put it in my pocket in the hope that I would see the robbers and would have a chance to attack them. Of course I put it back, but my rage, as my behaviour made only too obvious, had not gone away. One night's REM sleep was not enough to drain the emotional swamp, but over subsequent days the fury faded and I can now think of the experience with no real concern.

At the end of the night, the cancerous crocodiles paddle off and a neural dawn chorus begins. It arouses one part of the brain, one nerve transmitter and one hormone after another, until the whole system whirs into action. The process is slow and complicated and sometimes – as Proust and Paul McCartney both discovered – there is confusion as to what belongs in the land of dreams and what takes place in real life.

The excitements of darkness and the escape into daylight are very real to those who experience them, but they do not explain why such rhythms evolved in the first place. One obvious possibility is that the process conserves fuel. Compared with the need to sprint after a mammoth or a bus no doubt it does, but it saves no more than a tenth of the total burned up by a day in the office, and far less than that used by a farmer or a hunter. Perhaps just the brain saves energy, for in gentle slumber it uses half as much glucose as when awake. Rapid eye movement sleep makes it work harder, but that lasts for just a quarter of the night, so that overall the contents of the skull may indeed husband their resources as its owner falls into torpor. A night's sleep hence restores the exhaustions of the day; in Macbeth's words, it acts as 'sore labour's bath, balm of hurt minds, great nature's second course, chief nourisher in life's feast' (he had been busy that day, as he had just murdered Duncan, King of Scotland).

When awake, the brain is busy and generates a variety of unpleasant by-products. Another nocturnal task is to flush them away. A liquid called cerebrospinal fluid bathes and helps nourish the organ and also protects it from blows to the skull. The body makes about half a litre of the stuff each day. It circulates throughout the grey matter and in the tiny spaces that surround the nerve cells themselves. At night, the support cells of the brain, the glial cells, shrink by about half. That opens up those spaces, and increases the flow of the essential fluid. After it has rinsed the brain, it drains into the body's lymphatic system (the internal transport network). It does the job ten times better at night than in daylight.

Among the detritus washed away is a protein called beta-amyloid, an excess of which leads to Alzheimer's disease (and those who suffer from insomnia are at more risk of the condition, for just one night without slumber leads to a noticeable rise in the concentration of the compound).

The immune system, too, is also refreshed in the hours of darkness. The bone marrow, the source of white blood cells, is more active at night, which hints that this might be the best time to give a leukaemia patient a marrow transplant. The rate at which wounds heal also slows in the sleepless. Infectious diseases, too, respond. People with flu tend to feel tired, and may go to bed, where they nod off. Inconvenient as this may be, their somnolence is a sign that their defences are at work. Sleep is a powerful drug, so much so that people short of even three hours' rest are, when given a dose of cold virus, at twice the risk of infection as are those who get their full eight hours.

A lapse into oblivion also helps the body to repair the physical injuries it has suffered since it lurched out of bed hours earlier. Mutations happen all the time, but most are put right by a series of enzymes that repair them. In both mice and men the job is done with greater efficiency when asleep. Ultraviolet light is a potent agent of mutation, but most of the harm done is repaired by a specific enzyme. Mice irradiated in the morning develop five times as many skin tumours as those treated twelve hours later, as a further hint that restoration has a rhythm. Rats, too, when forced to stay awake for ten days show a real increase in DNA damage. After just two nights of decent rest, those mutations are repaired.

The gut, too, watches the clock. People who eat an identical meal for breakfast, lunch and dinner have a smaller spike in blood glucose after the morning repast than the evening meal, which means that, in effect, the body sees dinner as twice the size of breakfast. Even its vast population of bacteria undergo regular swings in composition as the hours tick by.

The ability to deal with other chemicals also changes as the day wears on. The breast-cancer drug tamoxifen works better if taken after breakfast rather than after dinner, while cisplatin, used to combat lung cancer but at the cost of unpleasant side-effects, is tolerated three times better when given late. For heart attacks, which tend to take place soon after waking up, treatment is often given early in the day. The painkillers morphine and oxycodone, like de Quincey's laudanum, themselves influence the circadian system, so much so that some addicts exchange day for night.

Other medical specialities have begun to take notice of the internal clock. Burns heal faster in people who experienced them in daylight hours rather than at night, and some claim that heart operations have a better outcome when they are done in the afternoon rather than, as is the usual habit, in the early morning. Hospitals, like prisons, tend to run to their own timetable. In intensive-care units, for many patients day and night are in effect abolished. They may be sedated throughout the twenty-four hours, fed on a drip both day and night, and rest motionless, sometimes for days or weeks, in constant artificial light. Their main internal clock may drift away from its normal rhythm, as – perhaps just as important – may the peripheral clocks that respond to regular mealtimes, to patterns of exercise, to external noise, and more. Although the evidence is not yet persuasive, some physicians now try to impose a regular pattern of light, of noise, of food and of therapy in the hope that it may improve their patients' prospects.

Dawn and dusk, it seems, pervade our bodies both in sickness and in health. Now, science has begun to disentangle how, and even why, they do so.

Edinburgh has, on the slopes above Princes Street, the world's oldest floral clock. The device is something of a fraud, for it is driven not by nature, but by electricity. Older versions were more true to the study of biological rhythms. In the eighteenth century, Linnaeus designed (but probably never built) a Horologium Florae, a circular bed filled with blossoms that would open at different hours to tell passers-by the time of day.

At about the same time, the French astronomer Jean-Jacques d'Ortous de Mairan developed an interest in the sensitive plant *Mimosa pudica*, which folds its leaves at night. He placed his subjects in a dark cupboard and noted that they continued their rhythm. He did not speculate on how they did it, except to say that they could 'sense the sun without seeing it'. Now we know that the plant has an internal chronometer. Like our own, it strays away from the truth if kept in constant light or dark, in a cycle that varies from around twenty hours to about thirty, and, like the animal version, it can be reset by exposure to light. He suggested also that humidity had an effect, and he was right, for to keep a plant in constant light but vary the amount of moisture in the air in a regular fashion also influences its behaviour.

Long after leaving Edinburgh, Charles Darwin became a martyr to his doctors, who could do little to help his episodes of insomnia and depression. One of his pastimes when laid up was to study the daily rhythms of shoots and flowers. When he noticed that a pot plant in his sickroom circled round as it grew, he fixed a glass plate above it and marked the position of its tip at intervals. A young shoot could make a complete revolution in just a few hours. Perhaps, he thought, it had an internal timer.

He also noted that the leaves of some species fold up at night. He describes this as sleep, but adds that 'hardly anyone supposes that there is any real analogy between the sleep of animals and that of plants'. He was too pessimistic, for many of the sensory systems of plants are related to our own. Plants do not have a brain, but to tell the time they do not need one. They have cells that sense the hue and intensity of light and share parts of their mechanism with animals. We have to make do with three such receptors, for red, blue and green, but plants have several more, which reach from the infrared into the ultraviolet. As we ourselves do, they gauge the time of day in part by assessing the amount of blue and red light, which alters as the hours move on. As is the case for animals, a burst of bright sunshine can stir a plant out of torpor, which means that growers can persuade chrysanthemums to burst into flower out of season with a dose of red light. Other species measure the length of the night rather than day, to the same end. Plants even get jet-lag, for a four o'clock plant transported from Sydney to London takes several days to return to its normal rhythm.

In 1962, the French geologist Michel Siffré spent eight weeks in almost constant darkness in an icy Alpine cave. The idea emerged at the height of the Cold War, when the French army became concerned about what might happen to those forced to live in a nuclear shelter, perhaps for weeks on end. Siffré continued a cyclical sequence of alertness and passivity, of meals, of body temperature and so on, but, almost isolated as he was from cues from the outside world, found that his internal patterns began to slip away from a twenty-four-hour sequence, with a tempo about half an hour longer than that of the heavens. Soon he fell utterly out of synchrony with the sun and lost count of the days, but once back in the bright lights of Paris he soon settled back into his natural cadences. His clock had a mind of its own, but one that was happy to listen to advice from the heavens when it got the chance.

In a mirror image of his experiences, subjects kept in constant daylight in an Antarctic summer also found it impossible to keep to time, for the melatonin peak moved forward every day. Without dawn or dusk to set it, their internal machinery lapsed into confusion.

A weakness in Siffré's research was that he was able to turn on a bright light for a short time when he awoke. His experiment has now been repeated with much dimmer bulbs controlled from outside. They show that a typical internal oscillation is in most cases no more than about ten minutes adrift from twenty-four hours, with some variation from person to person.

Where does the clockwork live? In the 1920s a mysterious disease called encephalitis lethargica affected millions of people across the world. It arises when the immune system attacks the body's nerve cells. Many of its victims died within a few weeks. Some among the survivors suffered a condition rather like Parkinson's disease and were frozen into immobility or filled their days with repetitive movements. Others had bursts of rage.

Post-mortems revealed that all the afflicted had an inflammation of the hypothalamus, at the base of the brain. The comatose and the wakeful patients had damage in different places. In his remarkable book *Awakenings*, Oliver Sacks describes the marvellous success in the 1960s of the drug L-DOPA (a chemical precursor of several nerve transmitters), which brought them back to the real world, for a time at least. Some of those raised from slumber had lost forty years of their lives and awoke, not with a memory of the previous day, but of one that had passed decades earlier.

The two forms of the illness hint that the brain, like many other parts of the body, has a system of checks and balances, with elements that say 'forward' matched with others that say 'back'. In healthy people they generate regular rhythms, but in the encephalitis patients just one takes control, to give years of unconsciousness or days of wakefulness.

The problem finds its source in a segment of the hypothalamus called the superchiasmatic nucleus. It has just twenty thousand nerve cells altogether, a tiny fraction of the brain's total, and is at its busiest in daylight. Its machinery is so robust that its cells retain their twenty-four-hour rhythm for weeks when kept in culture.

The suprachiasmatic nucleus acts as a Greenwich Time Signal that coordinates a series of secondary timepieces, on scales that operate from organs to tissues, cells, and individual genes. Even red blood cells – which have no DNA – can maintain a rhythm when isolated as a hint that part of the machinery may reside in the body of the cell rather than in the double helix. Such peripheral clocks bow to the rules of their master for most of the time, but listen for their own cues when they get a chance; thus, the regular arrival of food leads those in the intestine to change their activity to accord with mealtimes, whatever the sun might say.

In creatures such as worms or insects that cannot regulate their own metabolic rate, the central clock machinery is more or less resistant to changes in the thermometer, for if it were to run fast on hot days, and slow on cold, the whole system would break down. The same is true of our own central timer when its cells are grown at different temperatures in the laboratory.

That clock has several back-ups. They manifest their presence in people who have had strokes that destroy the superchiasmatic nucleus, for many of them still sleep and wake on a regular basis. One of the secondary timers responds to the drug methamphetamine. This plays havoc with its users' patterns of rest, and as 'crystal meth' has become a drug of addiction, in the United States most of all, where some of its devotees use it to fuel sex orgies that last for days without cease.

The network of receptors, hormones, electrical impulses and cells that lies behind the rhythm of life is intricate indeed, but is based on a simple principle. Each timer consists of a set of feedback loops that talk to each other and listen to external cues – light most of all – that modulate their activity.

The cyanobacteria, those survivors from three and a half billion years ago, were among the first to trap the sun's beams. Their chronometer uses just three proteins that interact in different combinations to adjust to the passage of night and day. When this simple system is introduced into a bacterium that has no internal rhythms of its own, it takes up the same behaviour. The timer also works for weeks when put into a test tube.

The clockwork gets more complicated as we ascend the evolutionary tree, but fruit-flies, mice, men and more share a whole series of genes that act as accelerators and brakes, some active soon after daybreak and some just before sunset.

The first to be discovered makes a protein that builds up over a night and is broken down in daylight. Errors in this *period* gene alter the length of the internal cycle, or abolish it altogether. To get the *period* substance into the nucleus from the main body of the cell, a second gene, *timeless*, is called upon. When that is damaged by mutation, the rhythm disappears even when *period* is still at work. Yet another element, *doubletime*, is also involved. Further components have emerged over the years, but the basic feedback mechanism applies to all of them. In flies and mice, several of them have variant forms that cause the mechanism to run fast or slow, or to lose all sense of rhythm. Similar changes may be involved in some of the differences in the way we regulate our own lives.

A few people rise with the dawn and go to bed when the rest of the world is still in front of the television, while others do the opposite. For the former, alertness and body temperature both peak in the middle of the afternoon, while for the others the high point is late in the day. Some people even retire at around six and rise at about two hours after midnight. They may try to change their ways, but the urge to slip between the sheets cannot, they say, be resisted. The condition has a simple pattern of inheritance, and mice engineered to carry the gene involved alter their behaviour in just the same way.

For most people simple habit is more important than biology in the decision as to when to wake up and when to go to bed. Benjamin Franklin ('Early to bed, and early to rise . . . ') began his day at five o'clock. He then sat around naked for an hour as he asked himself his invariable question: 'What good shall I do this day?' Thomas Edison and Ernest Hemingway, too, were among the larks. Marcel Proust, Oscar Wilde and Samuel Johnson were, in contrast, owls. Dr Johnson wrote: 'I have, all my life long, been lying till noon, yet I tell all young men, and tell them with great sincerity, that nobody who does not rise early will ever do any good' (a sentence that graces the title page of my own PhD thesis, for my supervisor used to complain about my late arrival, but never noticed my even more delayed departure).

Most of us lead lives more conventional than those of Johnson, Franklin, or even my student self, but about one person in six can be defined as an owl or a lark. Genes may be involved, but the main agent for owlish behaviour is a lack of bright light. In a simple experiment, a group of students at the University of Colorado at Boulder made up of individuals with a mixture of both habits were told to go up into the mountains and camp, with no artificial light at all. Within a week, all had converged on a melatonin rise at sunset. The bright sun they saw all day, and its contrast with a pitch-black night, reset their internal machinery almost at once, and soon they all went to bed, and arose, at the same time.

The contrasts between the Johnsons and the Franklins of this world are not small, with an eight-hour difference between the bedtimes of those at the extremes, a shift equivalent to that on a flight from London to Los Angeles. When Proust met James Joyce at a grand dinner in Paris in 1922 (the other guests included Pablo Picasso and Igor Stravinsky), the French author did not arrive until two hours after midnight. Joyce records that their chat was brief:

> Our talk consisted solely of the word 'No'. Proust asked me if I knew the Duc de So-and-So. I said, 'No.' Our hostess asked Proust if he had read such and such a piece of *Ulysses*. Proust said, 'No.' And so on. Of course, the situation was impossible. Proust's day was just beginning. Mine was at an end.

Proust was a snob of gigantic proportions (as Joyce further records of their conversation: 'Proust would only talk about duchesses while I was more concerned with their chambermaids'), although his diaries and letters reveal that he ameliorated his views with masochistic affairs with waiters at the Ritz.

Even so, and whatever his eccentricities, Proust was France's prophet of lost time. James Joyce was his Celtic equivalent. He wrote *Ulysses* a decade after he had left Dublin for the last time, but could still describe in minute detail its streets, its pubs and many of its characters (as he once put it: 'My head is full of pebbles and rubbish and broken matches and bits of glass picked up most everywhere'). For Joyce, as for Proust, smell was part of the patchwork of recollection. His experiences are less elegant than were those of his cross-Channel equivalent, with his recollection tweaked not by the scent of a delicate cake, but by the robust aroma of his hero Leopold Bloom's breakfast dish followed by its consequences: 'Most of all he liked grilled mutton kidneys which gave to his palate a fine tang of faintly scented urine' and, soon afterwards, in the outhouse: 'He read on, seated calm above his own rising smell.'

The first-century Roman orator Quintilian wrote of 'a curious fact, of which the reason is not obvious, that the interval of a single night will greatly increase the strength of the memory ... time itself, which is generally accounted one of the causes of forgetfulness, actually serves to strengthen the memory'. The idea has entered common parlance in phrases such as 'sleep on it', and is now at the centre of the science of slumber.

An observation made many years ago found that students asked to learn a series of nonsense syllables could recall them better after a night's rest than after the same period awake. That experiment has been replicated with various modifications again and again. The effects are associated more with the non-REM episodes, particularly in their deepest phases. They are not small, with improvements of a fifth or more in the ability to retain the information in those who spend eight hours in bed compared with those who stay awake for the same period. Slumber improves creativity as well as recollection. One researcher asked his subjects to work out sums and remember the result. After a night between the sheets they remembered the figures about ten per cent better than they did after the same interval in daylight. There was, in addition, a clue hidden in the answer, for the last three numbers of one sum gave, when reversed, the solution to the next. Once that had been noticed, the task could be completed with no difficulty. Twice as many of those tested after a decent night's rest spotted the trick as did the control groups.

It is even possible to learn while in the arms of Morpheus. The proof came when two groups of students were taught to play a pair of songs on a simple keyboard. Then they were allowed to nap for an hour and a half, and when they were asleep, one of the tunes was replayed many times to one group, and the second to the other. In each case, the subjects were better able to pick out the tune they had heard while they slept than the other one. Unfortunately, the many attempts to learn a poem, or a foreign language, by listening to a tape when in bed do not work, for these involve the accumulation of new information, rather than – as in this experiment – the recollection of material acquired when awake.

Whether the event recalled is the scent of a childhood delicacy in a French country house, or an argument with a friend in an Irish Martello Tower, sleep is the bank manager of lost time. It rejects some applicants as worthless, allows others to open a current account (but no more), and stores the contributions of a few in longer-term deposits, where they may remain for years. During the night's rest, the brain cements the day's experiences into the great edifice of recollection that makes each of us what we are.

Every second of every day we pick up something new. Most of the information leaves a trace for just an instant, with no mark on the conscious mind. Some is stored as soon as it arrives, but lasts for just a few minutes, hours or days. Other events can be recalled after decades. The information that streams in must be sorted into what might be useful and what is not.

Every recollection has a life of its own. I can remind myself what I did a minute ago, but will not retain that information for long. I can, on the other hand, recall the journey on the night bus to Scotland on the day I left home, and the less happy moment when I was attacked in Camden Town. Both events happened just once, but each has left a permanent imprint.

Other voices from the past, from French irregular verbs to the layout of a typewriter keyboard, needed more effort to lay down. Repetition is the soul of scholarship, and I look back with some nostalgia at my primary school days as we chanted the eleven times table and how to spell 'accommodation' and 'seizure'. I cannot, on the other hand, match the feat of Suresh Kumar Sharma, who once recited the first 70,030 digits of the mathematical constant 'pi' without a mistake.

Telephone numbers reveal some of the subtleties, and the challenges, of recall. Many phone systems can read out the number of the last person who called, but tend to do so in an uninflected monotone. That is a mistake, because most people cannot recollect more than about seven items for longer than a few seconds. Most British numbers have eleven digits, so that many of us cannot retain them all for long enough to return the call after hearing them just once.

One way to fix such numbers in the mind is to say them out loud several times. That holds on to the information for a little longer and hints at a separate kind of memory used when a message is spoken rather than heard. Another tactic is to group the information into blocks. My own home telephone is reachable at (with some editorial changes) 020 7946 0291. From the brain's point of view that seems to cut the number of separate items from eleven to three, which makes them easier to summon up on demand. Even then the information goes only into a short-term account, so that when the call has been returned, the digits are lost for ever.

I can, on the other hand, remember with perfect accuracy the telephone number of the Edinburgh flat I left half a century ago and can even summon up that of my wife's mobile. The bank of recollection has, it appears, many deposit boxes, and its accountants often move their contents from one to another.

Where might the sorting office be? It lies within the

hippocampus, a paired structure on either side of the brain, named after its supposed resemblance to the shape of a sea horse. Particularly in the young this produces large numbers of new nerve cells each day, a process once believed to be almost absent from the brain.

Hamsters, squirrels and rooks can store thousands of nuts and the like in hundreds of locations, and remember where they are for months. Their hippocampi are larger than those of related species that lack such talents. London black-cab drivers, too, are obliged to memorise the map of the city's twenty-five thousand streets, together with the locations of hundreds of hotels, hospitals and tourist sites. The task takes years, and candidates for the Public Carriage Office licence are familiar figures on their mopeds as they peer at a list of routes on a clipboard. I once asked a cab driver where the grave of Giro, the dog belonging to Leopold von Hoesch (the ambassador under the Nazis before he was replaced by Ribbentrop), might be, and he knew at once (it can be found next to the Duke of York steps, close to the Royal Society's offices, an edifice that was in the 1930s the German embassy). My question was too simple, he said, as the examiners preferred more elliptical queries such as 'Where is the only Nazi monument in London?'

Black-cab drivers as a group have, like squirrels, large hippocampi. This is not because those blessed with a capacious organ find the test easier, but because it grows as they study, and their recall of streets and sights improves to match. Those who fail the Public Carriage Office exam show less progress in either measure.

The hippocampus contains specialised cells that fire off when its owners pass through a familiar space. That remembered map makes it possible for cab drivers (and to a lesser extent the rest of us) to make routine journeys with almost no conscious effort. Other sections are specialised to deal with memories of words and sentences, or of ideas in general. The left hippocampus is more involved in language, and its opposite number in navigation. Women, some say, cannot read maps, while men prefer not to ask the way. True or not, males do activate their right hippocampus more than its partner, while females do the opposite.

Science can do more with rodents than with taxi drivers. Rats taught to swim in cloudy water to a platform hidden just below the surface soon learn where they should make for, because each successful discovery reinforces the memory link, but those whose hippocampal receptors are blocked by drugs paddle aimlessly around until they blunder into the target.

The most dramatic evidence for the role of that structure in memory emerged from an individual tragedy. Damage to one member of the hippocampal pair causes no more than minor problems, but when both are destroyed the results can be calamitous. In the 1950s, brain surgery was in its infancy. Lobotomy – the destruction of sections of the grey matter – was much used, even for conditions such as depression. Its inventor was even awarded the Nobel Prize for his misguided idea. After a bicycle accident, a young American student, Henry Molaison, suffered a severe and potentially fatal epilepsy. The removal of one hippocampus sometimes helped, but HM (as he was referred to in the scientific literature) had both the left and the right versions of the structure cut out. That controlled his fits, but almost destroyed his ability to recollect the past. He lost all awareness of every event in the year before the operation, even if he could drag up a few facts from the more distant past. Even worse, HM was left with no conscious ability to retain any new information, but saw, read, heard and lived in what his doctors described as 'a permanent present tense'. Henry Molaison cooperated with brain scientists for fifty years and became a celebrity in their world, although he was never aware of his fame. He died in 2008 at the age of eighty-two.

A three-dimensional atlas of his brain, based on thousands of sections, confused what had seemed a straightforward picture. Some of the hippocampus on each side had been left behind, but other parts of the grey matter had been damaged by the surgeon. Even so, his predicament provided several insights into the memory machine. Molaison's short-term recall system lasted for just a few minutes for each item and had remained in good shape, but he was unable to move any of that information into a more permanent store, as a signal that two distinct processes were involved. His ability to speak, and to do crossword puzzles (as long as they did not need information more recent than his surgery), also suggested that a stockpile of information picked up long ago was held in a separate compartment. Even after the operation, HM could, when asked to learn a simple task, improve his abilities (and in old age he learned to walk with the help of a frame), as an additional clue that to learn by rote involves a mechanism distinct from that used to recall a recent experience.

His plight emerged from groundless self-assurance, an affliction common among doctors of those days (and perhaps even some today). His surgeon, William Beecher Scoville – known, tellingly, to his colleagues as 'Wild Bill' – confessed later to a certain regret that his subject had suffered such a ruinous experience, and the story is still surrounded by controversy about the claimed destruction of data by a researcher after Mr Molaison's death.

De Quincey – who himself became famous for his ability to recall (or imagine) the past – thought that no memory was ever erased; that 'endless strata have covered up each other in forgetfulness'. The difficulties faced by HM suggest that he was wrong, and we are beginning to understand why. The bank of reminiscence does most of its business at night, when its manager sorts out high-quality deposits from low, saves the former and junks the latter. Such 'smart forgetting' begins as soon as the lights go out. The nerve junctions that transmit information from one to the next using chemical signals passed across a gap known as the synapse are busy during the day, but are scrutinised after sundown. Many of the links made ease apart in the hours of darkness if they fail the test of permanence. Transitory recollections of trivial events are quashed, while important information is allowed to survive. The sorting process readies the grey matter to face a wave of new deposits on the morrow.

Synapses that receive a constant and repetitive stimulus

from outside boost their affinity with their partner, while those not often used lose interest in each other. In the rat experiment, the brains of the drugged animals had failed to strengthen the links made between nerves each time the hidden platform was found, while the synapses of the untreated individuals were happy to stay in contact with each other after a series of successful returns to the place of safety.

This process is at the centre of both learning and its opposite. Slumber weakens the links between all synapses, and breaks many of them. In mice, for example, the area of contact decreases by around a quarter after a decent period of rest, with the most recent, and weakest, links most liable to abolition.

In both mice and men, memories judged as worthwhile are transferred from the current account into long-term deposits where a few may be locked away for years. Such 'memory consolidation' transfers information upwards in the brain, from the hippocampus to the cortex, the site of consciousness. Items from the previous day's diary begin that journey in the first part of the night, when slow waves dominate. REM sleep then allows a few of them into the vast reserve of facts stored there. Even among that privileged minority, many will sooner or later lose their value and will in days, weeks, months or years be replaced.

Perhaps that is how the need for sleep itself arose. Balancing the books of the past is hard work, and the bank within the skull is too busy to do the job in daylight, when it needs to deal with the queue of customers anxious to get in. Instead, the arithmetic takes place when its owner is between the sheets. There seems to be no particular reason why DNA repair, hormonal changes, the excretion of waste, or energy conservation (all of which happen at night) should demand a period of oblivion, but perhaps Dr Johnson's 'irresistible stupefaction' reflects the body's need to sort out the day's experiences in peace.

In belated homage to Proust, science now studies memories not just of faces and places, but of tastes and smells. The nervous pathways involved are distinct from those for sight and sound, because smell and taste are unique in their direct connection to the cortex, the home of long-term memory. All other senses are processed elsewhere before the information is passed to the brain. The loss of those two talents may be a first sign of the onset of Alzheimer's disease.

New-born babies can recall the scent of their mother, and will crawl towards her rather than to other women, and if she has eaten pungent foods such as peppermint or almonds while pregnant, her infant is more prepared than average to tolerate, or even enjoy, their odour. Everyone has their own memorable scents. I can go back to the time of the *Abbey Road* album with a whiff of patchouli, and to a decade earlier with the stench of the formalin that preserved the corpses of the dogfish I dissected in school.

Sleep, scent and recollection go together. In one experiment, students learned to recognise a series of images flashed up on a computer screen while they inhaled a scent of roses. They then went to bed. Some were exposed to the odour while they were between the sheets, while others were denied that experience. Those given the scent did better at recalling the images when they awoke than did those deprived of it. And, in a final (and somewhat desperate) attempt to put scientific flesh on literary bones, to expose a somnolent subject to the smell of a rose is said to raise the chance of a dream about the countryside.

Proust had anticipated such experiments long before. The cakes in *À la Recherche du Temps Perdu* were madeleines, and it was the scent of the ground almonds used to flavour them that took the author back to his infancy, when his aunt used to give him a piece of that confection when he visited her bedroom. As he writes: 'the taste and smell of things remain poised a long time, like souls, ready to remind us, waiting and hoping for their moment, amid the ruins of all the rest; and bear without flinching, in the tiny and almost impalpable droplet, the vast structure of recollection'. That sentence, recast in modern terms, hints that its author had more insight into the science of memory than he ever realised.

CHAPTER 7

BETWEEN DOG AND WOLF

When I lie down, I say, When shall I arise, and the night be gone? and I am full of tossings to and fro until the dawning of the day.

Job on his torments (Job 7:40)

Proust's hypochondria and dejection came in large measure from his decision to detach himself from the rays of the sun. For years he slept through the day and wrote through the night in a bedroom lined with cork to keep out the noise of the neighbours, with curtains always drawn and lit by a solitary green-shaded lamp. The cycle of the seasons made matters worse, for his fear of infection meant that he avoided the outside world even more than usual in winter. Time's chronicler often lay awake for hours, plagued by pain and by unease: 'my bedroom became the fixed point on which my melancholy and anxious thoughts were centred.' He became depressed and solitary. Soon he lost touch with nature's rhythms altogether and, perhaps not by coincidence, died in his cheerless, gloomy cell at the age of just fifty-one.

He was not alone in his darkness. Robert Louis Stevenson, once more:

> To none but those who have themselves suffered the thing in the body, can the gloom and depression of our Edinburgh winters be brought home. For some constitutions there is something almost physically disgusting in the bleak ugliness of easterly weather; the wind wearies, the sickly sky depresses them.

Today, millions of people experience sickly skies by choice rather than necessity. For many of us, time is out of joint. The young in particular have become people of the twilight, creatures 'entre chien et loup', the French phrase for the dusk in which they spend much of their existence as they peer into mobile phones, computer monitors and television screens.

A life in a real, or even a metaphorical, Edinburgh can still lead some to despair. Even in my time there, the Scottish capital was gloomier than today, for the ancient facades of the Royal Mile were then lit just by scattered incandescent bulbs which did little to brighten up the street. Stevenson had gaslights with even feebler beams to alleviate his dejection, although his writings hint that they did not do much to solve the problem.

The city's winters a century earlier were funereal indeed, with that thoroughfare blessed with just a few oil lamps. James Boswell was born beneath their flames, in Parliament Square, at the upper end of the Royal Mile, in 1740. From his youth he had an equivocal relationship with daylight.

He felt 'dreary as a dromedary' if forced to get up before noon, and often was 'frightened to lie down and sink into helplessness and forgetfulness'. His *Edinburgh Journals* tell of whole nights drinking, gambling and arguing with friends, even at the cost of attacks of despair over subsequent days. These were often intensified by guilt at his constant sexual dalliances and the infections that resulted: 'I unhappily went to the street, picked up a big fat whore, [and] lay with her upon a stone hewing in a mason's shed just by David Hume's house'. He made a resolution always to go home from those sessions in a sedan chair, safe from temptation, but did not hold to it.

Whatever the price of his dissipation, those evenings may have had one positive effect, for they involved the consumption of vast numbers of oysters, whose health-giving contents, Vitamin D included, may have done something to mitigate the effects of endless darkness. Visitors to the Oyster Club, founded by Adam Smith, the chemist Joseph Black and the geologist James Hutton, included the architect Robert Adam, David Hume, Benjamin Franklin and Boswell himself so that, in their own way, the mollusca did their bit to contribute to the Scottish Enlightenment as they had long before on the South African coast.

In fact, Edinburgh in the days of that august body was, thanks to its spanking new oil lamps, better lit than it ever had been. Just a century earlier, the nation's capital, its English equivalent and cities across the world had been plunged for half the hours of the year into a darkness now almost impossible to imagine. The boundaries between day and night were as unequivocal as they had been for most of history. At sunset, almost everyone went home, locked their doors for security, and stayed inside until dawn. Movement through the streets was almost impossible. Edinburgh had walls and gates that closed off large sections at night. The Royal Mile itself was blocked by the Netherbow Port, which was manned at all times but could only be used to enter between dawn and dusk. The entry fee was too much for the poor, so that many residents never left the city in their lifetimes. They referred to the gate as the World's End, for it marked the frontier of what they knew.

The curfew began at eight. In winter the darkness was relieved with feeble candle lanterns made of horn, with a narrow slit for the light, which citizens were obliged to hang outside until the bell tolled and blackness descended. Indoors, life was just as murky. Beeswax candles were beyond the reach of all but the rich. Those made of tallow were weak, smoky, and gave out a foul smell. The poorest used rush-lights, reeds soaked in fat skimmed off the top of a stew-pot. Some places turned to other expedients. The stormy petrel was so common in northern Scotland that locals killed them off to save on candles, with a wick thrust down the corpse's throat and its flame fuelled by the bird's own fat. Throughout these islands and beyond, just five hundred years ago the borders between light and dark were for most people as firm, or almost so, as they always had been.

Then, as oil lamps replaced dim lanterns (or dead seabirds), the boundaries began to leak. A trickle of twilight became a torrent, and then a flood that washed away patterns of activity that had followed the sun for thousands of years. Since Boswell's day, the level of artificial light in the developed world has gone up by ten thousand times, and the rate of change has speeded up, with a twenty-fold improvement in efficiency even in the present century. Although the proportion of the energy budget spent on lighting has, at about six per cent, been fairly stable since the days of candlelight, every time a brighter form of illumination has appeared, people have rushed to take it up. White LED bulbs with a marked blue tinge are now in fashion, but they are set to be overtaken in the near future by devices tuned to the light sensors in the eye.

Not everyone shared the thirst for photons. Calvinist Geneva insisted that the dark was reserved for prayer. As Rousseau put it: 'God does not agree with the use of lanterns.' Even so, in refined London and even in John Knox's obdurate Edinburgh the lust for light meant that the streets were soon modestly ablaze. By the mid-eighteenth century, London was blessed with five thousand oil lamps and its northern equivalent with several hundred. These were succeeded by gas mantles that gave ten times as much light, and soon the British capital had many thousands in the streets and many more in houses and offices. Edinburgh was not far behind and gained its first gaslights in 1819.

The narrator of *Lost Time* was a victim of their rays. The book's first paragraph describes his childhood fantasies as he awoke before falling back into slumber. The tale then darkens. It tells of an invalid who sees

> with glad relief a streak of daylight showing under his bedroom door. Oh, joy of joys! it is morning. The servants will be about in a minute: he can ring, and someone will come to look after him. The thought of being made comfortable gives him strength to endure his pain. He is certain he heard footsteps: they come nearer, and then die away. The ray of light beneath his door is extinguished. It is midnight; someone has turned out the gas; the last servant has gone to bed, and he must lie all night in agony with no one to bring him any help.

Thus the passage from day to night, from youth to age and from joy to despair; and thus an introduction to the modern age, in which millions live a life detached from daylight, and – like Proust and his avatar – pay the price.

The real assault on sunset began with Thomas Edison and his development of a practicable electric light bulb in 1879. It continues today. The Royal Mile now basks in a hard municipal glare, with many of its tenements, banks and churches further lit up by their owners. Their LED bulbs give an instant, and almost cost-free, simulacrum of daylight at the flick of a switch. Thousands more are planned for Scotland's capital, to give even the most modest suburb a torrent of photons as intense as that in the touristic centre.

All this means that darkness has almost disappeared from the city, as it has from much of the world. Artificial light is now stronger than that of the full moon and the stars in the great majority of inhabited places. In half of them the night sky is ten times brighter than its natural level, so that two-thirds of Europeans cannot see the Milky Way because of the artificial glow. Only one in a hundred lives under a completely dark sky.

Dazzling as Auld Reekie's night-time streets may have appeared to Boswell, or even to my undergraduate self, even the most modern artificial lights do little more than mitigate the gloom. The eye has a remarkable ability to delude its owners, so that its proprietors are almost unaware of the crepuscular conditions in which they choose to live. Most of today's street lights illuminate the pavement with no more than a thousandth of the amount that shines on it on a cloudless midsummer day (and at noon even in a gloomy Scottish December, the Royal Mile gets twice as much light as its high-tech bulbs provide at night). A winter living room gets twice as much per square metre as does its pavements, an office perhaps twice as much again, although the light that teenagers experience as they peer into a computer screen in their darkened bedrooms is, in contrast, even feebler than that in the street outside.

The digital world has reset the hands of the clock for millions of young people. In spite of claims that everyone spends less time in bed than they used to, a survey of adult habits over the past half-century suggests that this is not the case, even if their time-frame may have been displaced to rather later than before. For adolescents, in contrast, the effect is very real. Information from diaries that go back to the early twentieth century shows that on average British teenagers now sleep an hour a night less than they then did. Much of the change has taken place in the past thirty years – the age of electronics – as they have chosen to go to bed later and later even as they are obliged to wake at much the same time.

To make matters worse, life indoors has moved to the blue, dawn, end of the spectrum, as TV screens, computer monitors and LED lights emit beams rich in that wavelength. Some devotees wear goggles that block such rays as the hours wear on, while a few such devices are designed to reduce their output. Even so, most users accept the artificial sunrise that rises from their screens even as midnight strikes.

As time and tide, as perceived by man, cease to obey the instructions of nature, both body and mind have to face the consequences. Aretaeus of Cappadocia was a Greek physician of the first century AD. For one condition, 'lethargy' as he called it – depression, as modern doctors would say – he found a treatment: 'Lethargics are to be laid in the light, and exposed to the rays of the sun (for the disease is gloom).'

Aretaeus was the first to identify a link between the body clock and mental health. His observation holds to this day. Across the developed world, about one person in four suffers from depression or anxiety at some time, with half of all long-term sick leave due to the condition. The rate is rising fast, perhaps because of a shortage of daylight. Many of those affected receive no treatment, and the call for a cure is urgent. My colleague at UCL, Lewis Wolpert, who himself has suffered from the problem, says in his book *Malignant Sadness* that 'depression is to sadness what cancer is to normal cell division' as a reminder that the periods of lowered mood that affect many of us are trivial in comparison to those more seriously afflicted.

The Romans, like the Greeks, had a word for it: *acedia*. Chaucer uses its cognate in 'The Parson's Tale': 'accidie maketh hym Hevy, thoghtful, and wraw ... He Dooth alle thyng with anoy, and with wrawnesse, slaknesse, and excusacioun, and with Ydelnesse, and unlust' until at last the sufferer falls into the ultimate sin: 'wanhope, that is despeir of the Mercy of God'. Some of those terms for the plight of the sleepless might find new life today.

Robert Burton, in his own 1621 *Anatomy of Melancholy*, had an explanation for the condition, for he – like Robert Louis Stevenson – was convinced that the weather was to blame: 'if it be a turbulent, rough, cloudy, stormy weather, men are sad, lumpish and much dejected, angry, waspish, dull and melancholy'.

Burton, like Aretaeus, Chaucer and their contemporaries, had no choice but to hand over most of his life to nature's timekeeper, and if its rays were obscured, his mood took a turn for the worse. Today's ability to avoid them altogether means that dejection has become part of many more lives.

Proust's own sense of hopelessness began when he was a few years old. His father, a wealthy doctor, chided the lad for what he saw as a self-inflicted 'neurasthenia' that manifested itself in insomnia, asthma, headaches, slurred speech, and pains in the joints. He blamed such symptoms on his wife's excessive attention to their child, and wrote a book on the condition, *The Hygiene of Neurasthenics*, with his own son as an exemplar. Marcel himself met several of the major psychiatrists of his day, and as a young man spent more than a month in the Paris clinic of Dr Paul Sollier, then the world's greatest expert on memory, in an unsuccessful search for a cure.

The experience gave him the germ of an idea for his novel. In it many of the characters are medical men, almost all described as incompetent, lazy and dishonest; they visit a patient's deathbed just to pick up a fee, or fake a sick note to ensure that a rich friend can obtain fresh croissants while others starve. Medicine itself was 'a compendium of the successive and contradictory mistakes of medical practitioners', but Proust learned a great deal from it, for he describes conditions such as migraine, epilepsy, dementia, syphilis and stroke. The brain gets dozens of mentions, while neurasthenia (a condition now lost from the medical vocabulary) is also referred to. Proust was interested in science as much as in medicine, and once even referred to his ability to manipulate time as equivalent to that put forward in Einstein's theory of relativity.

The great French author, as his belated arrival at Paris's intellectual dinner party of the twentieth century demonstrates, was an owl's owl, a man who sometimes went for weeks without daylight. The science, if not the literature, of sleep-related disorders has made much progress since he closed his eyes for the last time. Now we know that people with such habits have a heightened risk of depression, use drugs and alcohol to excess, suffer from anxiety and face a real danger of suicide.

In Aretaeus' southern home, lethargics were – and are – quite rare, but for those who live closer to the poles, winter gloom – seasonal affective disorder or SAD as it has been called – affects millions, many of whom lock themselves away for several joyless months a year. In Alaska, one in ten citizens has December despond (although a few of its citizens face the problem in June instead). In Florida, the figure is one in a hundred.

Closer still to the lands of eternal darkness, the issue becomes a real challenge. A sixth-century account of the people of Scandinavia tells of how they went from joy to misery as long days gave way to uninterrupted blackness. Accounts of the heroic polar expeditions of the late-nineteenth and early-twentieth centuries also noted the dismal effects of months without light. The doctor on Robert Peary's first journey into North Greenland wrote of the men's experiences that: 'I can think of nothing more disheartening, more destructive to human energy, than this unbroken blackness of the long polar night'. At the other end of the Earth, a member of Otto Nordenskjöld's three-year Swedish Antarctic Expedition referred to 'depression and increased irritability . . . especially during the dark season'; and, as one early explorer wrote: 'The presence or absence of the sun is a much more important matter to us than the state of the thermometer.'

Almost by accident, a doctor on the 1897 Belgian Antarctic equivalent, the first to spend a whole (albeit unplanned) winter on the ice, rediscovered a treatment forgotten since the Greeks. It was the precursor of a therapy much used today: 'The best substitute for this absence of the sun is the direct rays of heat from an open fire. From an ordinary coal or wood fire this effect is wonderful.' Although the Belgians did not realise, it was the benefit of light rather than heat that did the job.

As everyone who lives north of the Scottish border is well aware, the rhythms of the year have an inexorable course. Plants flower in April and begin to lose their leaves in October. Birds lay eggs as the days grow longer, and many migrate southwards as the nights return. Such cycles are maintained for years even in the laboratory as proof of an internal clock set in months rather than hours.

Migratory birds kept in twelve-hour days are restless every six months or so as the urge to move builds up. Men and women, too, have cyclical patterns of behaviour. Suicide and violence peak in April and May, perhaps as a relic of the era in which long days intensified the battle for mates. Until not long ago there were seasons for sex itself. Across the whole northern hemisphere births peaked earlier in the year in places furthest from the equator, perhaps because that arrangement allows infants to spend their first days in months with the least stressful weather. Such cycles were strong in the century or so before the Second World War, with as much as a five-month difference in the peak time for births between Maine and Florida. In today's endless artificial summers they have faded away, as further proof that without an occasional nudge from the natural world, habits that have existed for millennia may disappear almost unnoticed.

Some claim that seasonal depression itself evolved because it gave a biological advantage, since those who stayed inside in winter because they felt too gloomy to venture outside had a better chance of survival than did their fellows who exposed themselves to a hostile, hungry and icy world. That idea is impossible to test today. Although we do not know how – or even whether – it evolved, medicine has learned a lot about the mechanics of sadness, and can, at least in some cases, even mitigate its symptoms.

As the nights draw in, or a teenager decides to spend long hours in a darkened room, the body's circadian timer, attentive to daylight as it always is, finds itself out of phase with its owner's patterns of activity. Someone with normal sleep has a six-hour gap between the rise in the hormone melatonin and the halfway point of his or her night's slumber. Many of those who suffer from melancholia delay that onset by several hours, so that the midpoint of their night's rest comes too soon after the substance has started its work. A few among them have a mutation in the melanopsin sensor that makes it harder for their timer to respond to the daily hint that it has become time to retire, but many more are too entranced by television, or Twitter, to listen to its advice.

Japan has hundreds of thousands of *hikikomori* ('those confined'), who peer at screens for many hours each day, their mental state so disturbed that they stay cut off from the outside world for years (and perhaps not by coincidence that country has a suicide rate three times higher than our own). To match that, the Western world has over the past few decades suffered an epidemic of ADHD. It has been blamed on genes, on allergies, on diet, or on difficult pregnancies, but some doctors suggest that many supposed cases reflect no more than chronic lack of sleep.

Those filled with misery in the depths of December often

turn for relief to a simulacrum of sunshine. A SAD light uses a bright white-light bulb, sometimes loaded to the blue end of the spectrum to mimic the rays of dawn. A half-hour exposure does improve the mood of some people. Light therapy is not a cure-all, but is as good a defence against seasonal depression as are most of the medicaments used against it. It even has some effect when the beam is directed into the ear, as a reminder of how little we know about our relationship with sunlight. As a measure of the real power of the solar input, even the strongest bulbs in regular use are no more than a tenth as bright as is a summer day.

Some of those who fall into winter gloom have an inborn tendency to that mindset that does not reveal itself until the days get short. Boswell wrote: 'I was born with a melancholy temperament. It is the temperament of our family,' and he had occasional attacks even in summer. Mental upsets of this kind change the lives of millions. Some are extreme, with, in the condition known as bipolar disorder, a swing from near-despair to one of near-mania. The despair tends to set in when the clocks go back, but the mania often reappears when they go forward again (and is sometimes referred to by psychiatrists as 'March madness').

Most people who experience the sadness of autumn gain at least some relief when spring arrives. Other unfortunates do not, and suffer persistent bouts of feeling 'sad, lumpish and much dejected', even under blue skies. Many lives have been shattered by that inability to respond to the signals of nature. Bright light, or medication, each help some in that predicament, but many remain unmoved by both.

Even so, for a proportion of patients, drugs given in conjunction with, rather than as an alternative to, light have immediate results, and can release people who have been trapped in a mental cage for years. Relief may last for months, and if the individual begins to sink back into their unfortunate state a further burst of real or artificial sunshine can improve mental outlook once more, as a hint that more than one part of their internal machinery is out of order, and that the two treatments find different targets.

New remedies offer further hope. Ketamine was invented as an anaesthetic, but is much abused by thrill-seekers, who refer to it as Special K. They speak of 'K-holes' in which they fall into a trance-like state and lose their sense of time and of self, so much so that they cannot interact with anyone and may imagine that they are dead. After reading of their experiences, I have decided to stick to white wine.

Quite unexpectedly, ketamine has also been found to be an effective and immediate treatment for some cases of severe depression, seasonal or otherwise. Just one dose administered as a nasal spray can lift a person out of that state within a few hours, with a remission that may last for weeks. Many of its natural equivalents, the endorphins included, are made in the skin when exposed to ultraviolet, which hints at a close link between the chemical treatment offered by ketamine and the effectiveness of real or artificial sunlight in doing the same.

The drug strikes at the brain's source of negative emotions and does so to almost miraculous effect. It acts on what has been called an 'anti-reward' centre, which inhibits the excessive production of substances such as serotonin that improve mood. If the suppressor of joy does its job too well, its owners will suffer. When the machinery is reset by ketamine, metaphorical sunlight re-enters their lives. Its powers, and the newly discovered hormonal talents of the skin when exposed to ultraviolet, also suggest that depressed people might gain from time spent on a sunbed as much as under a SAD light.

Sunbed addiction is not unknown, and ketamine too can be dangerous. Its molecules bind to the brain receptors for the opioid painkillers such as morphine, fentanyl and oxycodone, the last now at the centre of an American narcotics crisis. Because of the risk of addiction, ketamine has not yet been cleared for treatment of mental disorders on either side of the Atlantic (although some doctors do use it off-prescription). If it does in time become accepted, the spectre of despond may be pushed back a little further.

Even mild attacks of that sensation can interfere with sleep, and the problem feeds upon itself, as those who suffer from it fall further into gloom. The faith of Job was tested with torments that included the death of his children, but painful as that was, he bemoaned his inability to fall into slumber to the same degree.

The Glasgow physician A. W. Macfarlane in his 1890 book *Insomnia and its Therapeutics* listed more than a hundred causes, from hard beds to 'neurasthenia' – the vague condition that we now recognise as a mix of fatigue, anxiety and depression. His restless Scots also suffered from gout, noise, tapeworms, rickets and drink, all of which intensified the problem.

About one in three Britons and Americans – and, in homage to their great author, rather more among the French – face insomnia at some time in their lives. For them, the peaks and troughs of hormone secretion flatten out. The problem is at its worst among the poor, the old, the solitary and the female. Loud music and traffic, or light, heat and cold, can disturb a night's rest, while pain and anxiety may do the same. Around half of older women suffer from the problem to some degree. For authors, it can become almost an occupational hazard. Vladimir Nabokov noted a typical night's rest as what he described as 2+1+1+2+1, the figures being hours of sleep, and the '+' signs 'intervals of hopelessness and nervous urination'. The problem (urination not always included) was shared by Proust, Kafka, Dickens, Conrad, the Brontë sisters, Sylvia Plath, Shelley, Keats, Coleridge, De Quincey and Wordsworth, several of whom wrote poems on the topic.

Insomnia can also be sparked off by the onset of Alzheimer's disease (and a drop in melatonin levels may be one of the first signs of that), of Parkinsonism, of some cancers, and of heart disease, and it may be more than a coincidence that some of the medicaments used to treat such illnesses also disturb the daily patterns of activity. Schizophrenics, too, tend to have an uneasy relationship with the clock, for many go to bed after midnight and rise when the day is well advanced. As a hint of a direct link, some of the drugs used to treat the condition change the activity of clock genes.

Since my departure from Scottish gloom I have been more or less free of the problem – with the exception, that is, of the aftermath of a long flight. Many people have suffered the unpleasant effects of a loss of synchrony with our planet as it spins. Most of us get over jet-lag quite soon, often with the help of a deliberate attempt to bask in the sun's rays (which is why a Christmas trip from Sydney to Edinburgh takes longer to recover from than does one in the opposite direction). On arrival back home at perhaps noon, my brain – which has not yet cleared customs – still perceives it to be nine at night, which leads to a large mismatch between the body clock and that set by the sun. It can take a week or more to recover from jet-lag after such a trip. Some take a dose of melatonin to trick the body into slumber at the appropriate time in the arrival city. I have never tried that and prefer to find what photons I can.

Jet-lag also reveals another of the clock's hidden talents. On home territory it sounds the wake-up call about eight hours after bedtime, but after a change of time-zone the appetite for slumber builds up and the internal timers demand a hefty dose of repose before they are satisfied (my own record, after a trip back from Death Valley, is fifteen hours). Adenosine – the chemical that builds up in daylight and heightens the urge to sleep – is in part responsible.

Time dislocation, and its effects on health, also affect millions who have never been on an aeroplane. On weekdays the alarm clock forces them to rise before they have had enough rest. They then suffer from 'social jet-lag', and are forced to counter that with an extended stay between the sheets at the weekend. Seven out of ten Americans face the issue. On average, the breakfast peak in electricity use in that nation is an hour or so later on Saturdays and Sundays than on weekdays, which indicates how many people choose to catch up on their missed sleep. On this side of the Atlantic, hedonism can do the job, for many teenagers stay up late on Fridays and Saturdays, but are wrenched back to normal on Monday. Keen clubbers (and some of my students are, I suspect, among them) can experience a five-hour time difference between weekday and weekend bedtimes. In effect, such pleasure-seekers make a return trip to New York every seven days.

In a less punitive dislocation, every European and every North American suffers twice a year when the clocks go forward by an hour in springtime, and back again in the autumn. The process began during the First World War in both Britain and Germany as an expedient to increase production and save fuel (and in the second such conflict Britain took up Double Summer Time for the same reason). It was once almost universal across Europe and North America, but Russia, Iceland and Turkey have already abandoned it. Now there is a move to abolish it across the whole European Union after the discovery that some people take up to a month to recover from even that slight disruption in their daily rhythm.

One British worker in six abandons their natural timers not in the search for pleasure or through government decree, but because they are obliged to work in shifts, either always at night, or with a move between night and day every month or even every week. Their internal sensors, responsive as they are to regular patterns of mealtimes and the like, then fall into constant conflict with the cues sent by the sun. Many of the body's functions then begin to run out of control.

Testosterone is at its highest level in mid-morning, but for night workers the surge comes in the two hours before and after noon. That may be responsible for the elevated risk of prostate cancer that such men face. For similar hormonal reasons many of their female fellows have menstrual problems, while their chance of breast cancer goes up by one part in six for every decade of night shifts. In response to that, the Danish government now gives compensation to the shift workers who develop the disease. Those who work nights are also more likely to be obese, and their risk of heart attacks and stroke rises by half compared to the general population.

Such effects are so large that the World Health Organization has declared all forms of shift work as a probable carcinogen. Even jet-lag has its dangers, for aircrew are at greater risk of cancer and heart disease, perhaps in part for that reason, while a lifetime of night shifts much increases the chances of depression. To abandon the body clock in the interests of economics affects the health of millions, but its dangers have scarcely entered the public mind.

Even moderate lack of rest dulls the senses. The Chernobyl explosion, the *Exxon Valdez* oil spill, and innumerable train, car and lorry crashes have all been blamed on long and irregular hours of work and the exhaustion that results. In the United States with its enormous distances and poorly regulated transport industry, more people are killed by tired truck-drivers than by those drunk or drugged. The country has a fatality rate per million kilometres twice as high as in Britain. The law limits the number of hours worked, but many drivers break it, because they are paid by distance covered rather than by time spent behind the wheel. Their risk of dying on the job is twelve times greater than that of other manual workers. On rest days a typical driver stays in bed for an hour and a half longer than when working, as the body tries to balance its sleep accounts.

The medical profession faces the same problem. Until a decade ago, British medical trainees in the first years of their career were forced to go for long periods without sleep. In the United States, many still face an eighty-hour week. Doctors asked to work a thirty-hour shift make five times as many serious mistakes in diagnosis, stick needles into themselves twice as often, and double the risk of a car crash on the way home compared with those who keep a more sensible timetable. In Britain the European Working Time Directive limits the hours worked to forty-eight a week, with no continuous period of work more than thirteen hours long. In practice, many medical students ignore that. After several nights with just six hours of rest they make as many errors on tests of the ability to respond to a flashing light as after two whole days with no rest. One common problem is that a tired subject falls into an instant nap of a few seconds – which for a brain surgeon may not be a good idea.

When such disturbances go too far the contents of the skull may be thrown off balance altogether. Six years after his first subterranean experience the French scientist Michel Siffré relived the event with a much longer stay in a Texas cave. By the end of the ordeal his personal day veered from eighteen to fifty-two hours long and he stayed awake for up to forty hours at a time. About halfway through his six-month stay he sank into depression and began to think of suicide, but in time he regained his stability and managed to stay underground for the allotted period.

The ascetics and hermits of the primitive Church saw such experiences as a gateway to heaven. Slumber was the work of the Devil, as it interfered with man's duty to adore the Saviour. The faithful started their prayers at midnight and ended them twenty hours later. After a few weeks of this, novices were warned that they would 'hear noises, crashings, voices, and tormenting devilish screams', but at last these would be succeeded by 'the experience of true grace', a sense of movement, of immersion in a reservoir of faith and of an intimate communion with Christ. Many who attempted to achieve that blessed state fell foul instead of what they called the 'noonday demon', an evil spirit whose imagined presence made their days seem endless and caused them to doubt their faith, so much so that some killed themselves. Their distress was due to a shortage of slumber, rather than of spiritual insight.

As the witch-finders of sixteenth-century Scotland, their equivalents in the Inquisition, and a variety of military torturers all knew, delusions of this kind can be used to convict an enemy of heresy or terrorism, or simply as a refined punishment.

Twenty-five thousand Americans are held in 'supermax' prisons. In some, the aim is to remove all cues of time, using constant bright light or long periods of darkness, with silence or endless bedlam. Prisoners are sometimes allowed out for an hour's solitary exercise a day, or in some places for three hours once a week. Soon, the parts of the brain that process external information begin to show drastic changes and, in the tradition of the Christian visionaries, those involved have a suicide rate five times greater than that of the prison population as a whole.

Bodies as much as minds suffer when rest is denied. In a cruel experiment a Russian scientist kept puppies awake and active without respite. Within five days, all were dead. As she wrote: 'a puppy deprived of rest for three or four days presents a more pitiful appearance than one which has passed ten or fifteen days without food ... I became firmly convinced that sleep is more necessary to animals endowed with consciousness than even food.'

Perseus, the last ruler of ancient Macedonia, was captured by the Romans and, so legend has it, perished in prison after he was kept awake for weeks at a time. That story may be apocryphal, but has real equivalents. Fatal familial insomnia affects around a hundred people across the world. It involves a change in just one gene that alters the way in which a particular protein is folded. The aberrant molecule is then laid down in the brain. The first hint of trouble is that patients find that they cannot sleep for more than two or three hours a night. In time they stay awake for days at a time. Paranoia and hallucinations follow and most of those affected die within a couple of years. In a less dramatic version of the same problem those who average less than six hours' repose have an overall mortality rate about twelve points greater than those who manage a full night's rest.

In the United States, insomniacs take ten times as much sick leave as does the general population. Liver function, the manufacture of blood cells, insulin secretion, muscle tone, stress response, fat metabolism, food absorption, stem-cell development and more show strong circadian rhythms and suffer when these are disturbed. A lack of rest can hence lead to a host of conditions that seem unrelated, from liver disease to heart attacks, cancer and muscle loss. Poor sleepers face a real increase in the chance of death from heart disease, and a smaller, but still substantial, risk of a fatal stroke.

One common side-effect is that insulin levels drop and blood sugar runs out of control. The hormones that give the sense of satiety also lose some of their effectiveness. An irregular or delayed bedtime is as a result a better predictor of obesity among British children than is the number of sugary drinks they swallow (an impressive statistic given that the typical teenager pours down a bathtub full of the stuff each year).

There are many supposed cures for the condition, but few work as well as that prescribed by W. C. Fields: 'Get plenty of sleep.' Some are simple. People tend to drop off more readily in a cool room. Benjamin Franklin insisted on an open window and no blankets, while William Harvey – who discovered the circulation of the blood – would not get into bed until he was shivering with cold.

Other ideas are more subtle. Samuel Pepys, like many of his fellows, slept better if he shared a bed, and if his wife (or one of his several mistresses) was not available he was happy to do so with a male companion, chosen for his ability to ensure a restful chat that would ease him into slumber. In the days when sleep was seen as an aid to digestion, medical advice was to lay one's head on a large bolster to direct blood to the stomach, or even to nod off on a chair. Cooling liquids helped, with distilled cucumber juice a popular remedy. The belief that oblivion began as blood withdrew from the brain led to the suggestion that the vital fluid should be forced to leave with cold compresses to the head, or even with the whole body wrapped in wet sheets. Other experts recommended that those afflicted anoint their feet with dormouse fat, drink the gall of a castrated boar, or brush their teeth with dog earwax. Many people have private rituals that help, and Proust himself could not nod off unless he wore tight underpants fastened with a special pin.

There are plenty of modern equivalents of such nostrums. One recent volume entitled *This Book will Send You to Sleep* is, says the Preface, 'guaranteed to be devoid of excitement', and its contents live up to that, with sections on railway gauges and the political crisis in Belgium that began in 2007. Other inventions include pyjamas which claim to cool the body and earphones that generate 'neuroacoustic sounds' to fool the brain into the desired state. One gadget sends out a beep every three minutes for an hour after bedtime, in the hope that in time its wearers will be so exhausted that they will plunge into slumber. The desperate can turn to the Magnesphere, an expensive bed enclosure which claims to restore the body even if the eyes stay stubbornly open.

Many of those affected by insomnia seek solace in chemistry. Hemlock can work, but the period of rest might be rather longer than planned. Proust's favourite, chloral hydrate, is addictive, and itself lethal after even a small overdose. Van Gogh wrote to a friend about his own approach: 'I fight this insomnia by a very, very strong dose of camphor in my pillow and mattress, and if you ever can't sleep, I recommend this to you' – a practice that may have helped drive him to madness. Barbiturates then became popular. They do the job, at least for some people, but there too the safety limit between the effective and the fatal dose is narrow. A dose of melatonin at the appropriate time sometimes helps, and other compounds that interact with that and other hormones to fool the brain into somnolence are under investigation.

An excess, rather than a shortage, of sleep can also be a problem. Joe, the Fat Boy in *The Pickwick Papers*, is introduced into the narrative as he stands comatose in front of a door upon which a few moments earlier he had been banging. When faced with the angry occupant, he replies: 'Because master said I wasn't to leave off knocking till they opened the door, for fear I should go to sleep.'

The mythical Joe – 'Young Dropsy', 'Young Opium-Eater', 'Young Boa-Constrictor' – was modelled on a real boy, James Budden, son of an innkeeper in Chatham, where Charles Dickens spent part of his childhood. He, like many of those afflicted with gross obesity, laboured to draw breath. Such people may choke or even face heart failure. Blood oxygen then goes down, sleep centres go into action and they fall into immediate slumber. Obstructive sleep apnoea, as the condition is called, is unpleasant and dangerous. Its effects can be reduced with a mask that pumps in air, or by surgery.

Narcolepsy, a sudden and involuntary plunge into deep slumber, has a more direct tie with the circadian clock. The first evidence came from Dobermann dogs, some of which fall into a coma at any sudden excitement, even a new bowl of food, and lie motionless until the event is over. They are unable to make the hormone orexin, which gives the brain a wake-up call as the night draws to a close. The body then responds to any slight change in the level of stimulation with a plunge into oblivion.

The boy who appears in the first lines of *Lost Time* awakes to imagine that he is a character in the book he had read in bed, but is content with imagination alone. Some of those who share his problem go further, for they are not content with fantasies but put their dreams into action in a strange mixture of sleep and wakefulness.

Such behaviour may go some way to explain the once common belief in mediums, spirit-writing and the like. Most people dismiss their activities as simple fraud (Darwin, who disapproved of Alfred Russel Wallace's credence in them, was convinced that they were liars, and no doubt he was often correct), but in the early days of psychiatry it became clear that many of those who describe life in heaven or write messages from beyond the grave are in fact acting out fantasies while they are in effect asleep and dreaming, but are able to talk and to walk as if they were awake.

One delusion shared by many mediums is of being a reincarnation of a deceased (and usually distinguished) person, and acting in character for hours at a time. Perhaps the most famous case was that of a Hélène Smith who was active in late nineteenth-century Geneva, and was studied by the University's then professor of psychiatry, Théodore Flournoy. When in her trance-like state she claimed to be a fifteenth-century Indian princess called Simandi, or a reincarnation of Marie Antoinette. On other occasions she travelled to Mars, and learned to converse in the Martian language (which turned out to be unexpectedly similar to French). Tests of her sensitivity to pain, of her muscular strength, and of her tendency to suffer physical collapse while on Mars or in India suggest that she was in a paradoxical state of semi-sleep in which she firmly believed the events she was enacting. Her mother, her grandmother and her brother all showed similar behaviour, which hints perhaps that inheritance was involved. There is indeed now evidence that to have a somnambulant parent much increases the chances of such behaviour in children.

Flournoy published the results of his investigations in a book entitled *From India to the Planet Mars: A Study of a Case of Somnambulism with Glossolalia*. That greatly annoyed Hélène Smith and her supporters, who felt that she was being accused of lying, a claim she angrily denied. Flournoy responded in an article in the *Journal de Genève* that: 'The purpose of my work was not to show how to reveal a dishonest medium, but instead to show how the phenomena of a sincere medium can be explained through subliminal psychic processes not hitherto sufficiently considered.'

Now his finding is widely accepted and the delusions and

acts of those who suffer from the problem have been much studied. Somnambulism – sleepwalking – poses obvious dangers. About one in fifty adults indulges in it from time to time, while children do it twice as often. Those with the condition tend to be easier to arouse from normal sleep than average, and, when they are, to put their imaginings into action. Quite often, such behaviour is a side-effect of other psychiatric problems.

The idea has been much used by writers, with a dreaming Lady Macbeth attempting to scrub the blood of Duncan from her hands – 'Out, damned spot, out, I say!' – while in Wilkie Collins's 1868 *The Moonstone* the hero abstracts a valuable diamond when in a drug-induced stupor and spends months in search of the culprit. In Bellini's opera *La Sonnambula*, the heroine is found wandering and is wrongly accused of being unfaithful (she acts out the truth to music while perched in a trance on a dangerous bridge). Writers themselves have been inspired by such an experience. Robert Louis Stevenson's wife Fanny recorded that, 'In the small hours of one morning . . . I was awakened by cries of horror from Louis. Thinking he had a nightmare I awakened him. He said angrily: "Why did you wake me? I was dreaming a fine bogey tale."' The 'fine bogey tale' was of the physical transformation of an upright citizen into a vicious murderer, and was key to the plot of his book *Dr Jekyll and Mr Hyde*, the germ of which had come, long before, from reading James Hogg's account of a justified sinner.

Medicine recognises that condition and its relatives as 'parasomnias'. They come in various forms. Each reflects incomplete arousal from non-rapid-eye-movement sleep, usually early in the night. Other forms involve sleep eating, in which those affected go, quite unconsciously, to their kitchens and gorge themselves on an unlikely mix of foods, sometimes even cooking it first. Another variant involves them in sex acts, alone, or with a partner and with or without his or her consent, events of which, again, they have no memory when they awake. Some people suffer from sleep terrors in which a nightmare becomes real and, quite often, even as they scream and as their heart pounds, they cannot be consoled. In all these forms of the disorder, the subject has no, or little, sense of pain and hence may injure themselves or attack anyone who tries to help them. All these unpleasant experiences show – as Proust was well aware – that sleep and wakefulness are not mutually exclusive states of mind.

In the real world, such behaviour can end in tragedy. There have been deaths – sometimes classified mistakenly as suicide – when people fall out of a window or crash a car into a motorway bridge. Sometimes an episode ends in horrific events. In 2009 a Welshman strangled his wife of forty years while they spent the night in their camper van. The night had been interrupted by noisy revellers and he dreamed that he had been attacked by an intruder. When he awoke, he found his wife dead, and went straight to the police. At the trial, his defence produced evidence that he had been a lifelong somnambulist, and he was found not guilty.

That defence, insane automatism as the law calls it, in most cases was once dismissed as fraudulent. Now, with brain scans and the like, and secret filming of the person in custody while asleep, it has become more common, with acquittal the rule rather than the exception. However, only a very few cases have used such a plea and given that perhaps one person in a hundred experiences at least a moderate form of parasomnia at some time in their lives, the reluctance to use the defence may be misplaced. Whatever the legal argument, parasomnia in all its forms shows how often the boundaries of slumber blur with those of wakefulness.

Proust's account of his avatar's own disturbed sleep is less dramatic, for he does not act out his nocturnal fantasies, but uses them as a theme to connect his experiences into a narrative, linked by memory. Throughout his vast work he paints an elusive landscape of joy, despair, broken friendships and romance. The narrator falls for unsuitable and unfaithful women, brushes against homosexuals of both sexes, and becomes involved in endless snobberies and betrayals. Its climax is reached when, at a farewell party, the tangled net of affairs and intrigue comes together as the survivors of the tale, now aged and decrepit but unwilling to accept as much, reminisce about the past.

The book's first chapters make much of the beauties of the countryside around the hero's childhood home in Combray, and of the girls he admires from afar. At its close, Marcel finds himself adrift as an old man in the Bois de Boulogne in Paris, a place once at the centre of his existence. It now seems to him false and debased, peopled not by the elegant friends of his youth but by what he sees as an uncultured and ill-dressed mob. There, the narrator becomes conscious that he will never find in reality the pictures stored in his mind, because the world he knew has gone for ever. He at once resolves to solve the problem by writing a novel – the one that the exhausted reader has just completed – in which he will recall his life and his loves before they fade for ever. On the final page of his monumental work, his search for lost time and its memories has just begun. Today's science of sleep is in much the same position.

ENVOI

FOSSIL SUNSHINE

Edinburgh alone is splendid in its situation and buildings and would have even a more imposing and delightful effect if Arthur's Seat were crowned with thick woods and if the Pentland Hills could be converted into green pastures . . . and Leith-walk planted with vineyards!

William Hazlitt, *Notes of a Journey through France and Italy*, 1826

I do not often receive threatening letters from lawyers or abusive emails from the general public, let alone aggressive comments in the right-wing press, but a few years ago I suffered a minor spate of such things. They demanded retractions, mentioned damages, used foul language and were part of a brief campaign that raged (if that is the word) around my innocent head. They were the result of some comments I made about the presentation of science by the BBC.

In 2010 I was asked by the Corporation's then governing body, the BBC Trust, to write a report on its coverage of that topic. I interviewed more than a hundred people, watched and listened to many programmes, travelled to remote regions of the nation and helped to carry out an analysis of the treatment of science on all radio and television channels. My report, published a year later, praised much of the general output, even as it criticised an obsession with spectacular wildlife shots with no ecological context and the lack of attention to cell and molecular biology, the largest single area of science. Not many people noticed, although a few films on molecules and cells did follow and, perhaps for the first time, producers in the Science Unit had a brief conversation with their fellows in Natural History.

My comments on news coverage attracted rather more attention. The journalists were not happy with my complaints about their dependence on press releases rather than original sources, their remarkable ignorance of the huge electronic journal database that most scientists use almost every day, their own lack of scientific expertise, and the shortage of contact between radio and television news, but the real fuss, inside and outside the BBC, was about what I termed 'false balance'.

When it comes to politics the Corporation is obsessed with what it repeatedly refers to as 'due impartiality'. My many queries about quite what 'due' meant, and how it could be recognised when it turned up, never yielded an intelligible answer, but most of its news professionals took it to imply that when one political party's view was discussed it should be matched with an equivalent treatment of its opposite.

For politics that might seem reasonable, but for science it does not work, because many results are accepted – at least provisionally – by everyone who specialises in a particular field. In my own trade, evolution, I never respond to invitations from creationists to 'debate' the idea that the Earth is six thousand years old, because the claim is absurd. Those who believe it are practising theology rather than science, for no conceivable information could change their minds.

Parts of the BBC seemed not to have noticed. On radio news most of all, a frequent approach when dealing with scientific discoveries that might be seen as contentious was to interview an acknowledged expert and to give equal time to someone with opposed views held through personal conviction alone. I illustrated the problem with a somewhat flippant account of a mathematician who has found that two and two makes four. On the *Today* programme (the BBC radio breakfast news bulletin), the scientist would, in his dull and hesitant fashion, be matched with a plausible spokesperson from the Quinquenary Liberation Front who argued with flair and passion that the correct answer was five. At the end of the piece, the presenter would duly intone that 'Two and two is probably closer to four than five, but the controversy goes on.'

The real example I used was global warming. Does it exist, and can it be blamed on human activity? Truth does not depend on opinion polls, but ninety per cent of atmospheric scientists in 2011, the year of my report for the BBC, felt that the answer was 'Yes'. A survey issued six years later of ten thousand papers in the technical literature that expressed a view on the topic found that by then ninety-eight per cent agreed. Even so, in news coverage quite often a scientist was matched with a politician, an historian, an economist or a determined contrarian who argued – with no call for evidence – that the figures are wrong, those who publish them are corrupt, or, most often, that this is still a matter of controversy and the question should be left open. Some of those spokesmen (but perhaps not all) are fronts for the coal, oil and gas lobby. Ludicrous as their claims might be, the spectre of 'due' has been enough to persuade the BBC to give them airtime.

The fossil-fuel industry has long been aware of the problem of man-made climate change. Once, its scientists did research and published papers on the topic, and were happy to discuss it in public, but their employers' tactics then underwent a subtle shift. Thirty years ago, an American drought led to dust storms, heatwaves and deaths, and sparked off an outcry about the effects of pollution on the weather. Alarmed by this, the Exxon-Mobil oil company instructed its employees not to report the facts, but instead to 'emphasize the uncertain in scientific conclusion regarding the potential enhanced greenhouse effect'. It spent millions of dollars on a publicity campaign, with advertisements in the *New York Times*, the *Washington Post* and the *Wall Street Journal* laden with statements such as 'Let's face it: The science of climate change is too uncertain to mandate a plan of action that could plunge economies into turmoil'. On both sides of the Atlantic denial soon became a doctrine. For large parts of the media, it still is.

The shadowy organisations that back its campaigns use the logic employed by the tobacco industry in the days when it claimed that it was not certain whether cigarettes caused lung cancer. The drug-pushers knew sixty years ago that the evidence was conclusive, but as one put it in a rare moment of honesty: 'Doubt is our product'. They lied in their own defence with great success (if the premature death of millions can so be described), and the first successful legal cases did not take place until the 1990s. The oil men copied their tactics, and even employed some of the tobacco lobby's think-tanks, lawyers and corrupt journalists. Like them, they indulged (as do their successors) in straightforward untruths; in 1997 one oil executive claimed that 'the Earth is cooler today than it was twenty years ago' (the year turned out to be the hottest on record up to then).

Uncertainty is powerful stuff. The year after that statement, a London doctor claimed to find, on the basis of a study of twelve children, a link between the MMR triple vaccine against measles, mumps and rubella (German measles) and the chances of autism. This finding, later shown to be fraudulent, led to a drop in its acceptance by a fifth, with several outbreaks of disease. Surveys of the anti-vaccine activists showed that the main reason behind their belief was the sense that there was no scientific consensus on the issue. When groups of people, one of which showed general support for the procedure while the other was concerned about its possible dangers, were separately given evidence that the triple vaccine was definitely safe, that it had risks, or that there was no agreement on the topic, the last group were the most likely to feel that their own views, initially favourable or otherwise, were after all correct. To sow confusion, justified or otherwise, seems to reinforce, rather than reduce, prejudice ('even the experts can't agree'), so that the adversarial approach used by our national broadcaster in its search for objectivity may instead do the opposite.

A spokesman for the climate-change sceptics then went further, and told the House of Commons Select Committee on Science and Technology that: 'Some argue that free speech does not extend to misleading the public by making factually inaccurate statements. But it does.' Uncertainty, justified or not, is still a useful tool for those who peddle propaganda.

My suggestion was that the BBC should abandon such tactics when it came to the established science of climate change. Not everyone went along with the idea. The *Daily Express* had it that I was guilty of 'quasi-Stalinist thought-policing', the *Telegraph* saw the document as a 'sustained and brilliant rebuttal to the threadbare notion that our state broadcaster is in any way capable of being fair and balanced', while the *Daily Mail* preferred to call it 'a wickedly cynical strategy to promote a false belief'. The *Spectator*, too, was less than impressed. My views on the greenhouse gas issue, it informed its readers, represented a 'cherishably stupid, rude, fatuous, crabby, bigoted, ignorant, petulant, feeble, fallacious, dishonest and misleading argument'. Although I too had a thesaurus to hand, I did not respond.

Some months later I talked to the then Editor of the *Today* programme about the BBC's overall response, and his own lack thereof, to my report, and asked him why his programme had not changed its ways. His logic, he told me, was that if he went into a pub with a hundred people in it, one of the drinkers would not believe that humans had affected the climate, and that such an individual 'deserved his voice'. My response was that another customer might believe that mental illness was due to demonic possession. Did his views, too, merit ventilation? My argument, it seemed, was absurd.

Five years later, a prominent opponent of the notion of man-made climate change made a claim of such blatant inaccuracy about temperature change over the previous decade that, interviewed later on the *Today* programme, I described his followers as 'deniers'. The presenter – himself sometimes accused of undue scepticism about man-made global warming, but on this occasion with his impartiality duly bristling – said that this was a term from theology, rather than from science. He was wrong. I deny that the world is flat, because science shows it to be a sphere, and the term I used is appropriate for those who spread falsehoods about climate or anything else.

Ofcom, the BBC's regulator after the government abolished the BBC Trust in 2016, gave the Corporation a mild rap over the knuckles about the denialist's interview on the grounds that the information presented 'was not duly accurate'. That statement, blunted by the appearance of the familiar qualifier, had little effect. It took the Corporation two more years to produce an internal document which at last accepted the science of man-made warming. It then established a training course to persuade its journalists to follow evidence rather than opinion. Whether that makes a difference, time will tell.

The real problem is not the science of climate change but the politics. Its history is exemplified in the tale of the solar panels on the White House roof.

In 1979 the possibility that the Earth was heating up because of the accumulation of carbon dioxide came to public attention when, after the Arab oil-price rise of that year, the mathematical models that predicted a rise in global temperature and had long been known to scientists began to be discussed in the press.

A few months later, President Jimmy Carter placed thirty-two of the devices on the roof of his temporary home. As he did, he reminded the audience that:

> No one can ever embargo the sun or interrupt its delivery to us ... In the year 2000, the solar water heater behind me, which is being dedicated today, will still be here supplying cheap, efficient energy. A generation from now, this solar heater can either be a curiosity, a museum piece, an example of a road not taken, or it can be just a small part of one of the greatest and most exciting adventures ever undertaken by the American people: harnessing the power of the sun to enrich our lives as we move away from our crippling dependence on foreign oil.

The panels were of the rudimentary kind that heats water directly, but even so they generated enough energy to run a typical American dwelling; and, given that to provide hot water takes one-sixth of the nation's energy use, and that

China still installs thousands of such devices for household use each year, 'rudimentary' may not be the right word.

Then came President Reagan. He at once declared war on the environmental movement and cut funds for research while allowing more coal-mining on federal lands. The panels were removed, with the President referring to them as 'just a joke'. George W. Bush furtively put in a few in the grounds of 1600 Pennsylvania Avenue to heat its swimming pool, but their return to the roof itself had to wait until the election of President Obama.

There was in those years, whatever the political atmosphere, a growing acceptance by scientists at least that an eye should be kept on the climate. In 1988, almost a decade after the Carter speech, the Intergovernmental Panel on Climate Change, the IPCC, was established. Its remit was to 'provide the world with a clear scientific view on the current state of knowledge in climate change and its potential environmental and socio-economic impacts'. It set out to ask where the thermometer stands now, and how it might change in the future. It has produced a series of forecasts of the probable changes in temperature until the end of the present century. The most recent was in October 2018 and the next is due in 2022.

The final recognition of the need for urgent action on the global rise in the mercury came, or so it seemed, with the Paris Agreement of 2016, ratified as it was by almost every nation in the world. It determined that, from four years after that document, every effort should be made to ensure that the extent of atmospheric warming since, then at 1°C above the level just before the Industrial Revolution, should be held to a maximum of 2°C, and expressed the ambition to hold it down to 1.5°C.

Each signatory is asked to use its own tactics to do the job, in the hope that their joint attempts will lead to a global strategy. The scheme is lubricated by a hundred billion dollars a year of aid to the poorest countries and although it has no mandatory powers, it has at least focused attention on the need for action. Several nations have already responded to its advice.

The coronation of President Trump changed all that. Even before his election he had written that attempts to mitigate climate change are 'just an expensive way of making tree-huggers feel good', and since entering the White House he has tweeted and spoken more than a hundred times on the topic:

> Scientists have manipulated the data ... The concept of global warming was created by and for the Chinese in order to make U.S. manufacturing non-competitive ... It'll change back again ... Earth is cooling at a record pace ... The same old climate change (global warming) bullshit! ... Scientists have a political agenda ... Is our country still spending money on the GLOBAL WARMING HOAX?

Once in power, his attack started almost at once, with attempts to make cuts in the funds available to organisations involved in research on the topic, a ban on the use of the term 'climate change' by the Department of Energy, and an import tax on Chinese solar panels. He has given his support to uneconomic mining operations, invented a fuel called 'clean coal', dismissed the panel that reported on air pollution from power stations, promoted fracking, reversed Obama's ban on a pipeline across America from the Canadian tar sands, and announced an intention to freeze fuel-efficiency limits on cars and trucks (America's largest source of greenhouse gases). The document behind that last policy supports its case with the bizarre claim that because the world will warm by 4°C by the end of the present century in any event, the extra greenhouse gases produced by vehicles do not matter (a rationale often used by smokers, who argue that if they are going to die of lung cancer anyway they may as well have another cigarette). Not all his schemes have managed to get through the courts, and several states, California most of all, have plans to implement their own carbon reduction programmes, but on the federal level the direction of travel is clear.

The fossil-fuel barons saw their greatest triumph in 2017, when Trump announced his intention to withdraw the United States from the Paris Agreement and to make it the only country apart from Syria and Nicaragua not to adhere to it. A year later the newly elected President of Brazil – a nation until then a global leader in the climate cause – threatened to do the same. At a follow-up meeting on the Agreement in Poland a year after Trump's speech, part of his administration's agenda became clear: 'The United States . . . will not allow climate agreements to be used as a vehicle to redistribute wealth'.

The hope that the Polish conference would galvanise the world to act on the IPCC's report was dashed when the United States allied itself with Russia, Saudi Arabia and Kuwait in agreeing to 'note' rather than 'welcome' its recommendations. The meeting ended with many fine words, and repeated calls for action, but little clarity in how this could be achieved. In unwitting homage to his successor's acts, several of the original White House solar panels are now, as President Carter feared, museum pieces, one of them on loan to the Science Museum in London.

America's view of climate science is still split by politics. Just one in five conservative Republicans believe that the world is warming through human agency, while four times as many liberal Democrats agree with that statement. On this side of the Atlantic, at least when it comes to much of the press and to some politicians, the same is true. All the negative comments on the BBC Trust report came from right-wing newspapers and magazines, and several Conservative members of the House of Commons and of the Lords are active in opposing the idea that human activities are the cause of the rise in temperature. The public, however, once split on the reality of global warming, has moved not only to accepting the science but to a demand that the fossil-fuel companies should pay some of the bill for solving the problem.

The sceptics' main difficulty has been evidence, which in the end tends to win all arguments. In the face of decades of lies about tobacco, the number of male smokers in Britain has fallen from six out of every ten in my schooldays to a quarter of that figure today. Now, the temperature statistics have come home to roost.

The stream of data is rather overwhelming but deserves due emphasis. 2016 saw a new global high, and although in the following year the figure dropped a little, that was because of a decline in El Niño, the movement of warm water across the Pacific that pushes up global temperatures. Each of the last three decades has been successively warmer than any equivalent period for more than a century. 2016 was warmer than 2015, which in turn beat 2014, while 2018 is the fourth warmest year ever recorded.

In 2016 parts of northern Russia saw the mercury rise to 33°C above normal (and at times since then the North Pole has been warmer than London). The last dozen years have seen the lowest levels of Arctic sea ice since satellite measurements began. In 2017 a liquid gas tanker travelled through northern seas from Norway to Korea, and a few months later another made the journey in winter.

In the far north, the wave of warmth may have been welcome, but in other places it was less so. Basra, the southern port in Iraq, was once known as the Venice of the Middle East. The city has now become almost uninhabitable, with frequent dust storms and air temperatures occasionally approaching 55°C, close to the maximum in Death Valley.

Everywhere, the thermometer has reached new heights. A temperature of over 51°C was recorded in the Sahara Desert, the highest ever measured in Africa. Yerevan, the capital of Armenia, hit 42°C for the first time while the citizens of Tbilisi in Georgia, of Taiwan, and of Japan also saw the mercury rise to new levels.

As a more parochial sampler of the problem, in the months before this book went to press, a series of British records fell. Scotland saw its highest temperature ever, when Motherwell, near Glasgow, reached over 33°C (an unofficial figure, as the authorities were concerned that a hot car was too near the temperature screen), and Glasgow itself the hottest day ever experienced, at just under that figure. Belfast, too, broke through the previous ceiling. England faced its longest drought for almost half a century, and in July 2018 hospital visits to its Accident and Emergency Departments reached a height that surpassed those for the coldest months of winter.

Perhaps the most telling statistic of all is that the five years leading to the 2019 publication of this book are the warmest on record, and the twenty top seen in historic times have all been in the past twenty-two years. Given such figures – indeed, even given only that last-mentioned – almost everybody not determined to deceive themselves or the public now accepts that rapid climate change is real and that humankind is to blame.

When did all this start, and what is yet to come?

Various projections of its possible effects have been made. Twenty years ago they were crude and uncertain, but slowly they have improved. The global thermometer has risen by just over 1°C since the Industrial Revolution. By the end of the present century, a rise to 2°C above the early level would mean the depopulation of parts of the tropics because of lethal heatwaves. With 3°C (by no means out of the question) many cities on coasts (and eight of the world's ten largest conurbations fall into that category) would have to be abandoned at least in part, and food production would decline. Four degrees – the figure assumed by the Trump administration – means the spread of deserts in China and elsewhere and the disappearance of many Pacific islands, while large parts of many American coastal cities will be underwater. Five degrees seems inconceivable, but is not, and if reached would lead to cataclysm.

Much hot air has been expended on how to manage the problem, and some progress has been made, with a strong shift to renewable power in Britain and elsewhere, and an emphasis on improved fuel efficiency for cars and factories. Even so, efforts so far have been slow, painful and inadequate, with the global output of renewables still only one part in twenty-five of total power output.

The problem is almost entirely man-made. Plants and animals on land and sea have long poured out vast quantities of greenhouse gases. They still generate more than twenty times as much as does industry each year. All, or almost all, those emissions are soaked up by the natural forces that drive the carbon cycle. A single fifteen-centimetre cube of chalk – the compressed remains of tiny marine diatoms, their bodies filled with calcium carbonate – taken from the White Cliffs of Dover contains the equivalent of more than a thousand litres of carbon dioxide extracted from the atmosphere of seventy million years ago, when those creatures breathed their last. Such rocks of that kind are found fifty kilometres below the surface, where they have been dragged by moving continental plates, as evidence that life has been balancing the books of the carbon trade since the Great Oxygenation Event unleashed the dioxide of its element on the world.

The human contribution, in contrast, does not have a sink large enough to soak all of it up, so that the gas builds up in the atmosphere. That carbon overdraft has been mounting without cease since man first harnessed the flames to light a fire. A little of what rose from the Great Fire of London is still up in the air, as is much of that added since *On the Origin of Species* was published in 1859. Even since the IPCC was established three decades ago, more carbon has been released than was emitted in the interval between Darwin's noble work and the panel's first meeting.

Just before the invention of the steam engine, the air held about 280 parts per million of the gas. Much of it came from plants and animals in their breath, in their excreta, and in their corpses. Forest fires, domestic boilers, volcanoes and rotting vegetation also helped, as did the solid dose added by farmers as they cleared land for crops or cattle. As the smoke-stacks sprang up in Manchester, Brussels and Pittsburgh, by the 1920s the level reached three hundred units, a figure last seen three hundred thousand years earlier, before humans of modern form appeared on Earth. Now the air contains more of that compound than it has for three million years.

In 1958 a station to measure its concentration was set up at the observatory at the summit of Mauna Loa in Hawaii. The first sample gave a figure of 310 parts per million. By 1988 that had reached 359, in 2016 the level broke through

400, and as this book went to press it breached 412. Far from the amount emitted being held steady or cut as demanded by the Paris Agreement, 2018 saw – after a pause brought on by economic slowdown – a new peak. India had the biggest rise, followed by China and the United States, with the European level stable, after a decade of slow reductions. The reasons are obvious, with increased use of coal, oil and gas, and show no sign of going away.

Nobody really knows how much higher the level of carbon dioxide will go. For the Paris Agreement to stick to its targets, the amount should begin to stabilise almost at once, but it shows little signs of so doing. If promises made, but not yet fulfilled, are factored in, the high point may be reached in the late 2020s, but with no action to control the release of the gas the peak will not come for another half-century, at which date reserves of fuel begin to run out. Whatever is done, our planet will get warmer for several years from now because of the inertia built into the system, much of which involves the excess solar energy now stored in the oceans.

The IPCC has issued a series of forecasts of possible future changes since it was founded. Each has been more pessimistic than the last. The report published by the group in October 2018 reaches new levels of alarm and calls for immediate action. Ancient Greeks knew that the sun gods have a darker side, but now they are a greater threat than ever. They must be placated at once if the climate is not to run out of control.

Until not long ago, there was a tendency by many nations to regard the 1.5°C target as no more than a pious hope. The IPCC makes it clear that such a view was a considerable mistake. Once, it was estimated that if the promises made by the Paris signatories are kept, the temperature in the last year of the present century is likely to be between 2.5°C and 3°C warmer than it was in James Watt's day, and that the world could, with sufficient investment, cope with that. The latest report is less optimistic.

It suggests that even a 2°C rise will cause massive damage and that there is an imperative need to hold the increase to 1.5°C. The difference between the two might seem trivial, but it is not. At 2°C, almost all coral reefs (including the Great Barrier) would disappear, while for the lower figure one in ten should survive. An Arctic free of summer sea ice would happen once a decade for 2°C, but once a century for 1.5°C. For the higher temperature, one in three people would experience potentially lethal heatwaves, for the lower just one in seven. Sea levels would rise by ten centimetres more for the higher target. And some experts suggest that there may be a one in ten chance that the 1.5° target will be breached within only five years.

The threat is hence immediate, rather than – as many have assumed – distant enough not to cause real concern. Action, the forecasters say, is needed at once to stave off longer-term disaster. The coming decade will be decisive. To avoid the 2°C rise, greenhouse-gas emissions must be reduced by almost half from 2010 levels within a dozen years from now, and must almost be abolished by the middle of the present century. To achieve this, every nation must at once abandon all plans for proceeding with new coal-, oil- and gas-fired stations, even if they are already in the pipeline, for their design life will be at least a decade. The proportion of energy obtained from renewables – which has expanded in recent years – must rise from a quarter of the total now to more than half within ten years and three-quarters by the end of the century.

Half the children born in the developed world in the year of publication of this book will live until then. Unless things change, they will experience a new and much less comfortable way of life.

So far, the warnings have been much disregarded. In the three years since the Paris Agreement big banks and other financial institutions have invested almost half a billion dollars in the development of polluting power plants, much of which has not yet been spent. In the Americas coal-mining and oil extraction go on apace. In Europe, England is about to allow fracking, Norway has extended its search for oil into the Arctic Ocean, while Germany continues to mine dirty coal. Australia even plans to expand coal production, and in 2018 its deputy prime minister commented that his country would not be dictated to by 'some kind of a report' (the loss of the Great Barrier Reef does not seem to worry him).

The plan put forward by the IPCC – which itself has no coercive powers of any kind – has several components, some achievable with sufficient cooperation and cash, but others speculative at best. They involve massive reforestation and a huge jump in green energy, matched by cuts in carbon output. There is also talk of 'geo-engineering', the development of new technologies that will change the input of solar energy, either by blocking it, or by reflecting more back into space, not to speak of an ability to extract carbon dioxide from the air.

The IPCC tends to be conservative in its forecasts, and some have criticised it for not mentioning the danger of 'tipping points' at which some gradual change reaches a critical level that leads to disaster. Possible examples include the complete collapse of the ice sheets and the reversal of the Gulf Stream, which is already running more slowly than it has for over a millennium, as the Arctic warms, cold fresh water pours off the Greenland ice sheet, and the jet stream in the upper atmosphere changes its route. Everyone is familiar with the comforting idea of Gaia, the sense that the Earth has a series of checks and balances that keep the climate in equilibrium, but, say some experts, we may now be entering an era of anti-Gaia, in which feedback loops turn positive, making a bad situation even worse.

How did we get into this mess, and how can we escape from it? And, if we do not, what will be the effect? The forecasts are gloomy but they are, after all, forecasts. A glance over the shoulder may improve their credibility. As Winston Churchill put it, 'The longer you can look back, the farther you can see forward.' Technology means that we can now look back at the changes that have taken place over the past few decades, over the centuries and over far longer periods, to put the present era into context. Science and history then combine to show that change is happening at unprecedented speed.

The study of the Earth's thermal balance began in post-revolutionary France, when the physicist Joseph Fourier worked out to his surprise that, given its distance from its own star and the force of solar radiation, our planet should be an icy desert. Something had kept it warm, and he speculated that the atmosphere might trap the long-wave radiation emitted by the surface after it has been heated by the sun. He had discovered the greenhouse effect.

Then came the British physicist John Tyndall. He used a device that transformed heat into electrical signals to study the capacity of various gases to block the infrared radiation given off by warm objects. In 1859 he measured the ability of oxygen, nitrogen, methane, carbon dioxide and other gases to soak up radiant energy. He soon found that water vapour is by far the most important insulator. Carbon dioxide, too, held back such radiation. The air acted as 'a blanket more necessary to the vegetable life of England than clothing is to man. Remove for a single summer-night the aqueous vapour from the air . . . and the sun would rise upon an island held fast in the iron grip of frost.'

He announced his results at a session of the Royal Institution, with Prince Albert in the chair, and Karl Marx an avid listener. He later repeated his account at the Royal Society, where it impressed Alfred, Lord Tennyson, who was in the audience. Edward Lear, who was also there, was less enthralled; he heard 'the diabolical Professor Tindal rave on gases, figures & the deuce knows what. At the end of an hour I felt I was going actually mad, & flew the brutal torture' (and as a schoolboy I often shared his sensations). Diabolical or not, Tyndall was right, and a Planet Earth that lacked the panes of the greenhouse would have an average temperature of 17°C below zero.

There are other gaseous villains in the tale, some familiar and some less so. The atmosphere up to a height of eighty kilometres or so is composed of seventy-eight per cent nitrogen, twenty-one per cent oxygen, just under one per cent of argon and 'trace amounts' of other compounds. Those traces have become a major cause of concern.

In Tyndall's day, carbon dioxide made up less than three hundred parts per million of the atmosphere, and it seemed to him that its role in the greenhouse effect could be ignored because water vapour was, he calculated, sixteen thousand times more significant. He did not notice the crucial difference between them. Water vapour stays aloft for just a few days before it falls as rain, while the gas can remain for centuries and builds up as more is released.

Tyndall was cautious about his results. He accepted that changes in the amounts of greenhouse gas might have given rise to 'all the mutations of climate which the researches of geologists reveal' and might indeed affect the weather in the future, but was not ready to warn the public about the dangers to come. He would have less reason for reticence today.

Forty years after his lecture, with the Industrial Revolution in full spate, the Swedish chemist Svante Arrhenius suggested that if factory chimneys caused levels of the compound to go up, air temperature would rise at a rate close to the square root of its concentration (an idea far too simple, but the first of many attempts to make a mathematical model of future climate). Even so, he felt that any change in the foreseeable future would be too limited to be noticed. He, like Tyndall, was wrong.

Carbon dioxide, the main greenhouse gas, has long circulated through air, land and sea. Forests and fields as they grow soak up some of it, but the sea is the most important storehouse, for it contains fifty times the amount held in the atmosphere. Since the start of the Industrial Revolution, some three hundred and fifty billion tons of carbon has been pumped out by human efforts, almost half of it since the foundation of the IPCC. Much of the gas has been taken up by those two reservoirs, but sooner or later it will return. Without Neptune's assistance, the greenhouse effect would have been obvious long ago, but his efforts, mighty as they are, have not been enough to stop the build-up in the skies.

Before the Industrial Revolution world production of coal was a few tens of thousand tons a year. British mining has been abandoned to such an extent that just before this book was published the nation's generators now have at least one coal-free month a year. Elsewhere the picture is quite different, for global production stands at eight hundred million tons, and shows no sign of decline.

The oil industry took off in the mid-nineteenth century with the Pennsylvania Oil Rush. Since then there has been massive expansion, with an annual consumption of almost five trillion litres. In 2006 the United States imported more than half its oil, but within little more than a decade the country had become the world's biggest producer. Natural gas, too, is in a phase of rapid expansion, with fracking set to generate still more, and there too the USA is now the largest source. The peak in output has led to a collapse in the price of both fuels, and in spite of all the pious words, their emissions continue to rise.

Five centuries ago, most of the Fens of East Anglia and Cambridgeshire were under water. Over the years they were drained, to reveal a flat, deep and fertile soil. In the 1850s, the largest remaining lake, Whittlesey Mere, was emptied. Its proprietor had noticed that areas drained in earlier times had begun to sink, and decided to measure the rate at which his own new fields did so. In 1851 he drove a cast-iron shaft deep into the surface until its base rested on the solid clay below. The top of that marker was then just above the ground. It now stands 4 metres high.

Where has the landscape gone? In the first couple of decades it sank fast as the ground dried out. Then the rate slowed, but it did not stop. The reason is both subtle and sinister. When the lake was still there, and in the first years after it was drained, the water had kept oxygen out of the soil, which stagnated and was built up by plant roots, bacteria and more. As it dried out and was exposed to the atmosphere, insects, worms, fungi and bacteria, soon joined by farmers, burrowed or ploughed through in the search for food. The fossil carbon reacted with the oxygen that got in to make carbon dioxide, which blew away. What had taken ten thousand years to lay down was lost to the air in a century and a half.

Mangrove swamps, peat bogs and marshy plains cover about one part in sixteen of the land surface and store twice as much carbon as do all the world's forests. They have been destroyed more rapidly than any other habitat. Ireland, in homage to the Celtic Mist, half a century ago generated almost half its power from stations fuelled by peat, which is even more polluting than is coal. It still has three, but they are scheduled for closure and there are attempts to allow the bogs stripped to feed them to regenerate. Even so, many marshlands across the world have suffered the fate of the English Fens and more are poised to follow. Draining the swamp, useful political slogan as it might be, has in practical terms turned out to be a toxic idea.

Forests have long been burned for farmland, and the flames are spreading. In Indonesia, they can spread to the peat below, sometimes to a depth of ten metres. In 2015 such blazes produced a pall of smoke that reached Malaysia and Singapore, with the release of several hundred million tons of carbon dioxide and a hundred thousand deaths.

Carbon dioxide gets the headlines, but the crisis has other culprits. Methane is less abundant than its fellow, and lasts only about ten years before it is broken down by ultraviolet light, but makes a solid contribution to the greenhouse. It has one carbon with four hydrogens, each attached to its central partner by a single chemical bond (carbon dioxide has each of its two oxygens linked to their partner element by a doubled-up bond). The infrared waves are soaked up faster when they hit a molecule that vibrates in sympathy, and single bonds do so more than do double. When measured over a century or so, methane traps long-wave radiation with almost thirty times the efficiency of its more familiar cousin.

Once, most of it was generated by bacteria and the like in places low in oxygen, such as marshlands, peat bogs and tundra. Methane also seeps out of the soil in regions underlain by coal and oil, is pushed out of volcanoes, and bubbles out of the permafrost and from Arctic lakes.

Man helps it on its way. The waterlogged soil of rice paddies generates lots of the compound. Reservoirs and rubbish tips also leak the gas, while coal and oil add to the damage as they are extracted and burned. Fracking, too, can release large amounts, for natural gas is itself almost all methane. In many gas and oil fields the excess is flared off from tall chimneys or simply pours into the air.

Our own production pales in comparison with that of farm animals. The inside of a sheep, a goat or a cow is a wild and woolly place. As those creatures digest grass and transform it into flesh, milk or wool they make an unconscious contribution to climate change. Each has a sort of pre-stomach into which they pass their food. There it is fermented and in part digested by bacteria, yeast, fungi and other single-celled organisms that thrive in the absence of oxygen. The animal then 'ruminates': it returns that mixture to its mouth, where it chews the semi-solid mass and swallows the blend into another specialised chamber, where fermentation continues, breaking the raw material down into products that can be easily absorbed (we do much the same, except that we use a saucepan rather than a spare stomach to predigest our food).

As the tiny ovine or bovine assistants do their work, they generate large amounts of methane, which emerge from one end or the other of their hosts' guts. The dose generated by a single cow in a year has the same effect on climate as does the exhaust of a car driven ten thousand miles. With half as many ruminants in the world as people, the threat from global flatulence has become explosive. Altogether, methane generates about a sixth of all greenhouse gases, with its concentration now at a level not seen for hundreds of thousands of years.

Other compounds such as nitrous oxide (otherwise known as laughing gas) also play a part. The amount in the air is a thousand times less than that of carbon dioxide, but the gas is three hundred times better at trapping outgoing radiation than is carbon dioxide. About a third of the atmospheric total is due to human activities. Some comes from industry and some from forest fires, but the largest source is agriculture. The eructations of farm animals do quite a lot of damage, but their solid wastes do much more (as, for that matter, do our own as they pass through a sewage works). Cattle, pigs and chickens are now raised in vast feed-lots or animal sheds which accumulate huge amounts of nitrogen-rich excreta. Satellites show that several times more of that gas and its relatives is being produced than assumed until recently, with hotspots in the Ganges plain in India and around the Yellow River, each a site of intensive agriculture. The factories which produce billions of tons of nitrogenous fertiliser also generate large amounts of the gas as waste. Given the essential and growing role of such industrial-scale farming, nitrous oxide may be even harder to control than is carbon dioxide.

Water, in contrast, seems innocuous, but it does its bit to add to the discomfort blanket. Look out of an aircraft window and, quite often, somewhere in the sky will be the trail of another plane. Such contrails soon seem to disappear, but they do not. I noticed the effect years ago, when

I worked in Death Valley. Fruit-flies like an early start, and we were up before dawn and out with dustbins filled with rotten tomatoes for bait as the desert sun sparkled in a bright blue sky.

Then, as the hours passed, it began to fade. Dozens of planes had passed overhead from Los Angeles Airport. Each left a white trail, which spread out as the day went on. In time, they joined to make a sheet of ice crystals, sometimes just visible as a thin and wispy cloud. Sunbathers may complain, but the shield's main effect is to boost the greenhouse effect as it traps long-wave radiation on the way out.

The world has pushed up the thermometer in other ways. Cities tend to be less reflective places than was the countryside upon which they were built. Dark roofs and roads have a lower albedo than do fields, while the loss of much Arctic ice together with the retreat of glaciers has replaced a shiny surface with dark rocks or waters that absorb, rather than reflect, the solar input.

Alpine glaciers began to shrink fast in the middle of the nineteenth century in a period that was not unduly warm. Ice cores show that at just that time there was a spike in the deposition of black soot on the mountains as the Industrial Revolution gathered speed. The besmirched surface then soaked up more of the sun's rays. The Himalayas now face the same fate. In 2014 a shift in wind direction brought so much black dust from China that within a few weeks the thickness of ice on one glacier dropped by a fifth.

All this has had an effect on the climate. Thermometer records for England go back to 1772. The rest of the world was slow to catch up, but by the middle of the nineteenth century enough stations were in operation to give a rough picture of global temperature. Today, thousands of instruments are scattered over every continent and are complemented by satellites, ocean sensors, and more. The information they provide can then be used to calibrate other measures – proxies, as they are known – that also track shifts in the level of the mercury over that period. The proxies can in turn be utilised to probe deep into a time before that device was invented.

In the shorter term, tree rings and their equivalents in corals, holes drilled in the ground to test how much it has warmed, and pollen grains trapped in sediments that hint at local ecology and climate all tell part of the tale. Another approach is to count stomata – the gateways that allow gases into and out of leaves. When carbon dioxide level is low, as in the thin air of high mountains, leaves have more of those pores. Centuries-old specimens in museums show general decline in the numbers in their descendants as a signal that there has indeed been a rise in greenhouse-gas concentration over that time.

Proxies over a longer period include the advance and retreat of glaciers, and the rise and fall of seas. Bubbles of air trapped in cores drilled deep into Arctic or Antarctic ice also provide valuable information. Oxygen comes in two forms with different atomic weights, O^{18} and O^{16}. The lighter variant tends to evaporate more from the sea in warm weather, so that the relative concentration of each in air trapped in such bubbles hints at what the temperature must have been.

Other records use the shells of foraminifera and diatoms. Their shells are in most cases based on calcium carbonate. They are found all the way back to the Cambrian era, some five hundred million years ago. Because the bottom of the ocean is not much disturbed, they stay in order and are of interest to oil companies, as well as to those concerned with the damage done by their products. Once again, animals that died in warm periods had more of the heavier version of oxygen than of the light.

Each of those techniques has strengths and limitations, but together they produce a record of summers and winters long gone. They are supported, with more or less credibility, by tales of vines in Labrador, of the Thames and the Bosporus blocked by ice, and of the collapse of ancient empires when crops failed after years of cold. They combine to show that we live in a unique moment in climate history.

The general picture from the birth of Christ to that of James Watt is one of a series of swings from warm to cold and back over a few decades or centuries. Ice cores from Greenland and Antarctica reveal dozens of spikes in volcanic ash and sulphur, each the footprint of an eruption that blocked the rays of the sun. In AD 535 and 536 several events took place, perhaps in what is now El Salvador. They were followed four years later by a further large explosion in the tropics. After the American eruption, global temperature dropped by about 2°C, and, after its successor, in some places by the same again. That marked the opening salvo of the 'Late Antique Little Ice Age', whose existence had long been suspected from the records and which lasted for a century and a half. One Byzantine historian noted 'a most dread portent ... the Sun gave forth its light without brightness'. On the other side of the world, the Emperor of Japan agreed: 'Yellow gold and ten thousand strings of cash cannot cure hunger. What avails a thousand boxes of pearls to him who is hungry and cold?' The population of Scandinavia collapsed, and dynasties from Byzantium to China trembled as cold, starvation and disease gnawed at their foundations.

As the dust settled and the clouds melted away there came a period of more favourable weather, the Late Medieval Warm Period. It lasted for around three hundred years from AD 950 and peaked at the time of the Battle of Hastings, an era in which vines grew in England.

The 1257 eruption of the Samalas volcano in the Malay Archipelago brought those happy days to an end. It was the greatest eruption of the past two thousand years. Lava travelled for thirty kilometres and a huge plume of sulphurous smoke and ash reached into the air. It blocked the sun's rays to such an extent that global temperature was reduced by 2°C. In London twelve months later, a mass burial took place in Spitalfields, perhaps as evidence of widespread starvation. A chronicler of the time described the disaster: 'Innumerable multitudes of poor people died, and their bodies were found lying all about swollen from want ... In London alone fifteen thousand of the poor perished; in England and elsewhere thousands died.' The capital had no more than fifty thousand inhabitants at the time.

Over the next two centuries there were smaller outbursts in Indonesia and elsewhere. Together they loaded the atmosphere with so much dust that a long episode of cool weather set in, the trigger for the Little Ice Age that lasted from the sixteenth to the nineteenth century (and enabled the French Republican armies that invaded the Netherlands in 1794 to capture the Dutch fleet, frozen into its harbour, in a cavalry charge). The final such event, at Tambora in Indonesia in 1815, brought Europe's and North America's 'year without a summer' (and led, incidentally, to Mary Shelley's *Frankenstein*, when the weather caused her and her companions Lord Byron, her husband Percy Shelley and the physician and literary figure John Polidori to stay in their Swiss villa and write ghost stories rather than take healthy walks).

Then the picture changed, with a rise in temperature that has continued until today. The climate sensors of Victorian times were not as sensitive as their modern equivalents and did not register the shift until the beginning of the last century, but hints of a grand warming can be picked up by proxies as long ago as the 1830s. Levels of greenhouse gas had by then gone up by just a fraction since James Watt, as evidence that even a slight rise can have an effect. Today, its malign influence is universal.

Climate has changed many times over the millennia, with no need for human intervention. Each period, be it an ice age or a warm interval, lasted for thousands or tens of thousands of years, with icy periods far longer than the warm intervals that separate them. The transitions are, in a geological context, rapid. Air bubbles in Arctic and Antarctic ice show that during the present climate cycle the level of atmospheric carbon dioxide in the cold phases is much lower than that in warm. Should the planet cool too far, ice begins to spread. To match that, after an upturn in temperature in an ice age the frozen shroud will collapse. Whatever the cause, the case for a repeated history of stability followed by change is persuasive.

Antarctic cores suggest that over the past three-quarters of a million years there have been about eight glacial advances and retreats, with a fairly regular period of around a hundred thousand years. The last advance began around a hundred and twenty thousand years ago and faded around eleven thousand years before the present.

At its peak most of Europe and North America was buried under a kilometre-thick winding sheet. Sea levels fell by one hundred and twenty metres and Britain became part of Europe. Then the thermometer crept up, the ice fell back, the English Channel opened, and we were on our own.

Such natural fluctuations have been blamed on many things. They include changes in the intensity of the sun's output and of volcanic activity. Variation in surface albedo and in levels of greenhouse gases as forests flourish or collapse also play a part, but the prime mover of climate is the uneasy relationship of the Earth with its path around the sun.

Our planet, like many of those who live on it, undergoes repeated bouts of eccentricity and obliquity. With a regular beat of just under a hundred thousand years, the shape of its orbit – its eccentricity – changes, from elliptical towards almost circular. The tilt of its axis from the vertical – its obliquity – cycles with a period of about half that. In addition its axis of rotation changes its inclination with the vertical with a slower rhythm. All three influence the amount of energy that comes in. They are the pacemakers of climate.

The Earth teeters on as it has done since it was born, and its temperature staggers up and down in synchrony, but in the context of human lives such changes have usually been too slow to notice. Those in the last geological instant – well within the lifetime of many readers of this book – most certainly are. They have already affected their lives and those of the creatures around them, and will continue to do so.

Extinction is the watchword of the twenty-first century. The numbers of wild mammals have gone down by a half in the past seventy years, with the present rate of loss a thousand times greater than that before the Industrial Revolution, with one species in thirty across a wide variety of creatures now at risk. Habitat destruction, in the tropics most of all, is much involved, but a survey of extinction rates across the world since 1950 shows that the best predictor is the rate of increase in temperature.

In 2007 I wrote a book, *Coral*, which set out to bring up to date Charles Darwin's first scientific work, itself published in 1842. In it he explained the then baffling observation that remote atolls were perched on tall pinnacles of dead coral – whose builders need sunlight to survive – that rose from the darkness of the deeps. As he realised, this came from the slow collapse of a volcanic island into the depths of the Earth, while the polyps that form the reef grew fast enough on the summit to stay in the sunlit zone.

My own rather less original volume had a subtitle, *A Pessimist in Paradise*, chosen to emphasise the dire state of the reefs at the time I wrote it. Events in the brief interval since it was published add a certain weight to that phrase.

The average surface water temperature around the Barrier Reef had risen by just under a degree in the time from Charles Darwin's visit to Australia in 1836 to the date of my own book. Since then it has risen much faster, with a peak of an additional three-quarters of a degree in 2016. Carbon dioxide dissolves in seawater, which becomes more acid; bad news for reefs, as their limestone foundations may be etched away. There has been an increase in acidity by half since the voyage of HMS *Beagle* and many of those structures have paid that chemical price.

They also suffer from the direct effects of climate change. In 1997 much of the Great Barrier Reef experienced a severe bleaching event. Such crises take place when the water gets too warm and the green algae that feed the polyps with the products of photosynthesis flee – or are expelled – and live free instead. Some reefs die, but most get over such traumas within a few years.

In the southern summers of 2015 and 2016 its temperature reached new heights, and a massive bleach took place in the northern and central sections of the Reef. A second disaster of this magnitude had not been expected to happen for at least thirty years from the 1997 event. Almost all the northern section has now perished and may take decades to recover, if it ever does. Climate scientists often speculate about the collapse of an ecosystem. In tropical seas, the process is well under way, and as their paradise is lost, pessimism has gained new vigour.

Coral reefs find it hard to escape from the challenges that climate throws at them. However, many other plants and animals, on land and in the oceans, can avoid the worst and are on the move. I was somewhat startled when I went into my local French supermarket a couple of years ago to see a stack of spray cans labelled with 'This kills tigers!' I was not aware of a local problem with big cats, but a second glance showed that the target was the tiger mosquito.

This creature, named for the stripes on its body, is native to tropical south-east Asia. It appeared in Albania in the late 1970s and then made its way to southern Italy. In France it reached Nice soon after the turn of the century and found its way to my department, the Hérault, within just a few years. Now, dozens of departments, almost all in the south, have been colonised, and there are pockets of the insect in the Paris suburbs and a few individuals have even been trapped in England. It can carry diseases such as the Zika virus, dangerous for unborn babies, and dengue, which can be lethal.

When they get a chance, even trees can travel at some speed. The rivers of ice at Glacier Bay in Alaska have withdrawn by almost a hundred kilometres since they were first seen in the eighteenth century. Fossil pollen shows that spruce and pine have moved northwards by fifty kilometres a century since then, and it can take little more than a hundred years to turn what had been an ice sheet into a mature forest. The birds, mammals and insects that live in it follow on.

The same process is at work in the ocean. Caribbean fish have been caught in British waters while blue-fin tuna, once rare, have become more common, even as the native cod and haddock are moving to chillier and deeper waters; and for the first time sperm whales have been seen in the Canadian Arctic.

Homo sapiens, too, is not immune from climate-driven migration. The Vikings were forced out of some of their territories with the onset of the Little Ice Age, and moved onwards to the British Isles, among many other places. The area in which I grew up, the Wirral Peninsula, can, to add to its many delights, boast of Viking village names: one, Thingwall, refers to a discussion group or 'Thing', Britain's first parliament; another, Thurstaston, to Thor's Stone, a red sandstone outcrop overlooking the Dee. There are even claims, based on the DNA of people who bear local surnames, that Viking genes still persist there. The millions now moving northwards into Europe and the United States are also driven by climate change, with heat and drought in Colombia damaging the coffee crop and driving farmers off the land, while the situation in Somalia is even worse.

Other creatures have changed not the geography of their lives but the rhythms. For a thousand years from AD 800 the average date at which the cherry trees in Kyoto, the ancient capital of Japan, burst into flower was the fifteenth of May. In the past two centuries that date has fallen back by two weeks.

Britain's plants have been tracked since 1736, when Robert Marsham began to study them on his Norfolk estate. He noted the dates that leaves opened in sycamore, birch, beech, ash and oak. He gathered the figures until his death in his early sixties, and his descendants continued the observations for almost two centuries. Thousands of volunteers now take note of the date of 'budburst' (the moment when the green colour of the new leaves is first visible through the scales of the bud) across the British Isles. Old photographs taken in the same locations suggest that spring now arrives around two weeks earlier, and autumn waits a week longer, than when I was at school. Because chlorophyll fluoresces in sunlight, satellites can track its arrival across the planet. On both land and sea, over most of the globe, in the past thirty years springtime has come earlier by around three days a decade.

Many of these changes have taken place within my own lifetime and, whatever the prospects for the famous Agreement, the inertia in the system means that for at least the next decade and perhaps longer, the world is set to get even warmer, with many experts predicting a three-degree change. How will that affect our lives, and what will happen if politicians ignore the urgent warnings of the latest IPCC report and the Paris deal collapses?

Perhaps, some say, the Apocalypse will arrive. Its four horsemen, of death, pestilence, famine and war, will gallop through the streets, brandishing thermometers. Lethal temperatures, the spread of tropical diseases, crop failure and conflict over food or water are, to the pessimists, certain to break out. Without action, they might well; but the time to save ourselves, as the IPCC point out, is now.

The most obvious danger is heat itself. If climate change proceeds unchecked, the wave of mortality seen in France in the heatwave of 2003 would rocket up, as the temperature there in a few decades might rise on some days to 50°C. Across the world about a third of the total population is already exposed for at least twenty days a year to heatwaves that can kill. That figure that may rise by half.

In terms of wet-bulb temperature – the joint measure of temperature and humidity – where the lethal level is 35°C, and danger level is around 31°C, India, Pakistan and China regularly experience 28°C and sometimes nudge 30°C. Without urgent action, a quarter of a century from now may see those figures move up by 2°C, to bring hundreds of thousands, perhaps millions, of extra deaths. The people of the North Chinese Plain, from Shanghai to Beijing, whose moist atmosphere is made worse by vast amounts of water used for irrigation, may face the greatest problem. Even in Britain, with no attempt to control the climate, by the end of the present century the coldest summer months will be hotter than the warmest of today and heatwaves of a severity expected to occur just twice a century at the end of the 1900s will arrive twice a decade.

For the first time, more than half the population of the world now lives in cities, which act as heat islands, which means that those who live in such places face twice the increase in temperature since the millennium than do those who live in the countryside. A study in 2017 of almost five hundred such places across the globe showed that an extra hundred and fifty million vulnerable people – the elderly, the infants and the unwell – were at risk of heat stress than seventeen years earlier; and that Europe had more of a problem than did the developing world because of its ageing urban population. Heat also increases the dangers of air pollution, which means that almost no cities in the developing world have safe air.

Maize is the main cash crop of the Western world. It evolved in the cool air of the Andes, and grows best at around 28°C. Its main centre is in the United States Corn Belt, with summers that sometime exceed this by twelve degrees. Although selective breeding, pesticides and new machinery have improved productivity, yield is sometimes little more than half what it would be if the crop were raised at its optimal, cooler, temperature. The same rules apply to oats, wheat, rice and barley. Without action on greenhouse gases, by the end of the century just half of today's seed crops will be produced.

Those who toil in fields or factories themselves flag when the thermometer shoots up. All over the world, production reaches a maximum at 15°C. As the mercury rises above, or falls below, that level, productivity drops by one per cent for every degree. That simple equation has held true for half a century. If temperature continues to rise, the output of labour will slow in decades.

The French political writer Charles-Louis Montesquieu's 1748 book *The Spirit of the Laws* is a landmark in legal history, but its author also suggests that 'cold air constringes the extremities of the external fibres of the body' and that those exposed to it were more vigorous, more courageous, more cunning and more polite than the effete inhabitants of the tropics. The ideal climate was, by coincidence, close to that of France.

Montesquieu may, in part, have been right. Within the United States a heatwave is followed, nine months later, by a drop in the birth rate, as a hint that sex and sticky nights do not go together. In school, the marks on arithmetic tests drop by several per cent, and drivers sound their horns more often, people are ruder to shopkeepers, and they use more foul language when online (on Facebook, the number of swearwords goes up tenfold on the hottest compared with the coldest days of the year).

All over the world the rate of violent crime goes up as the mercury rises. In Chicago, Milwaukee, Baltimore and Detroit, the number of shootings doubles on days with a temperature above 30°C compared with those below 10°C.

Crime waves can be managed, but those of the oceans are harder to control. A warmer world will see the landscape shrink as sea-levels rise – a fact of some interest to the six hundred million coast-dwellers who are at risk. Much of the effect is due to the newly melted ice of glaciers and of ice sheets.

John Tyndall, the discoverer of greenhouse gases, was a keen alpinist and student of glaciers. Much of his work concentrated on the Mer de Glace, which had 'the appearance of a sea, which after it had been tossed by a storm, had stiffened into rest'. In his day it stretched from Mont Blanc almost to the town of Chamonix. It has now retreated by two and a half kilometres, with a third of the decline since the turn of the present century. The same retreat has taken place in almost every mountain range in the world.

In truth, the world's mountain glaciers are minor players

in the ocean's profit-and-loss account compared to the ice-caps around each pole. Greenland is at present the biggest single depositor. Its melting rate has gone up by half since the Industrial Revolution with almost all that increase taking place in the last thirty years. Now, it is accelerating fast, in part because its newly wet surface allows algae to grow, darkening the ice and allowing it to soak up more solar energy. The newly liberated water of that empty island is responsible for about a fifth of the global rise in sea levels. The Antarctic will soon surpass it. There, a message for the future lies deep in the past. The rapidity of the rise in sea level at the end of distant ice ages suggests that its ice sheet melted far more rapidly than once supposed, even though the rate of warming was much slower than in the past century. That hints that its retreat may, after a tipping point is reached, feed on itself; perhaps because when a thick floating extension of the main ice cap breaks off it leaves a cliff of frozen water too tall to support its own weight. It then collapses, and the process repeats itself. If such feedback sets in as the world warms further, it might double present estimates of sea-level rise. Its contribution has gone up by three times in the past five years, and the potential for disaster is very real, for 2017 saw the highest seawater temperatures recorded, helping it to nibble away below the waterline.

At the end of the last ice age world coastlines were a hundred and twenty metres lower than today, but then the tides rose fast. From seven thousand years ago the tides slowed, and for most of the time the increases were of less than half a metre over several centuries. The dawn of the twentieth century saw a new and persistent surge, and by the millennium the tides had risen by about fifteen centimetres, at an annual rate of just under two millimetres. Then the figure went up by about a millimetre and a half a year, and there are recent signs of further acceleration, with the present rate at almost four millimetres a year. Melting ice plays the larger part, but expansion as the seas warm also has a role.

The overall rise in the oceans by the end of the century may be as much as sixty-five centimetres, although there are local peaks and troughs of increase, with as much as twelve millimetres a year off Japan, matched with regions, most of the Mediterranean included, where the level is almost static. Such differences come from local variations in salinity, geology and prevailing winds.

How far will the waves reach? Projections for the end of the century vary widely, from thirty to eighty centimetres, with a great deal of uncertainty in the estimate, which is very sensitive to slight shifts in temperature, and shows considerable geographical variation that emerges from changes in local circumstances. Should the whole of the West Antarctic ice sheet be lost the flood would rise to three and a half metres. That is not likely to happen within the present century, but is a herald of real threats to come.

Even if the 2°C target is achieved, by the end of the century sixty million Chinese will find themselves below high-tide level, and will have to move, or drown. The same is true for twenty million Vietnamese and twelve million Americans (not to speak of the entire population of the

Maldives, where whole islands will be submerged and its government will have to build new ones).

Nowhere in the developed world are the challenges more obvious than in Florida and on the shores of the Gulf of Mexico. The Galveston Flood of 1900 was the worst natural disaster in American history, with eight thousand deaths. The city was built, like today's Miami Beach, on an offshore island. Each reaches its high point less than three metres above sea level. The surge of water sucked up by the intense low pressure of a major hurricane covered it with two and a half metres of water.

Now the area's citizens are better prepared, with evacuation warnings, forced abandonment of coastal towns and new sea walls and levees. In spite of such precautions, nine of the ten most expensive marine floods in the past twenty years have taken place there.

The Gulf coast's problems are exacerbated because much of its landscape is sinking as it recovers from the rebound upwards at the end of the last ice age, and as oil, gas and fresh water are pumped from below. The effects of Hurricane Katrina in 2005 were exacerbated because the ground in some places is slumping at five centimetres a year. After the great storm, tens of thousands of African-Americans – who lived in low-lying areas of the city – fled New Orleans, never to return, and the process may just have begun. Within fifty years the southern states might lose a tenth of their population, and the north-east see a matching upturn. Elsewhere, the problem is more spectacular. Parts of Jakarta, the capital of Indonesia and a coastal city ten million strong, are, because of the extraction of ground water, sinking at two metres a decade, and at this rate a large section will soon be underwater.

Even on a moderately pessimistic forecast, within thirty years more than three hundred thousand homes in the United States will face a flood every spring tide, despite a fortune being spent on short-term measures such as the 'Big U Wall', a ten-mile-long series of berms five metres high to be built around Lower Manhattan.

Such problems can be seen, on a lesser scale, closer to home. These islands have still not recovered from the last ice age, when glaciers covered their northern half. When they melted, Scotland and northern England rose to the occasion, and the surface there is set to add a further ten centimetres by the end of the century. South-east England, in contrast, continues to subside. When I was a boy I was told that 'When the lions drink, London will sink'. The mooring rings on the Embankment are in the form of lions' heads, and when the Thames rises to cover them, any further increase would fill the streets. That moment may come again. The Thames Barrier opened in 1983. It shut its gates just four times in the 1980s, but in the year of the millennium it did so four times as often. It should do its job at least until 2030, but one day it will have to be replaced. As much as a third of the English low-lying coast, including Tilbury, parts of East Anglia and Lincolnshire, the Somerset Levels, and around Blackpool may have to be abandoned by the end of the present century.

Climate change has altered the world in many ways and

will continue to do so. What, as Lenin put it, is to be done? As in his day, experts disagree about the way forward.

The most obvious approach is to make a drastic cut in the production of greenhouse gases, with wind farms, solar panels, insulation schemes, electric cars and the like. Some progress has already been made, with the cost of solar power dropping to a fifth of its previous level over the past decade, with three million electric cars on the road.

Another view has it that the Climate Change Panel is not radical enough and that it would be quite feasible to reach a carbon-free economy with the technology available today. Wind, water and sunshine could generate all the power needed by industry, transport, households and farming by the middle of the present century. Everything would run on electricity or on hydrogen produced with its help. Demand for energy would drop by half and the lives of the five million people a year who now die from air pollution would be saved.

Progress has already been made to such ends. Norway, with its abundant hydroelectric power, is already a third of the way to green power; the United Kingdom is now sixth in the world for wind, with Germany ahead of us and France slightly behind, and it will soon shoot up in the rankings as the Hornsea One wind-farm off the Yorkshire coast, the largest in the world with turbines as tall as London's Gherkin office block, comes on stream to power a million homes. Every nation has its own specialisations. France is top of the list for tidal power, while Spain accounts for almost half the world's concentrated solar systems, in which mirrors or lenses focus the sun's heat onto a steam turbine that generates electricity.

Supporters of the alternative plan set out by the IPCC insist that its ideas are more pragmatic, more flexible, and less wedded to a single approach than such a purely ecological strategy. As well as investment in renewables the panel recommends that by 2050 all cars – if they exist at all – should be electric, and that public transport should be widespread and as cheap as possible. Meat consumption should be slashed and the throw-away culture should be abandoned in favour of long-lasting clothes, houses and appliances. The constant urge for economic growth must be resisted, and money should be poured into education, for a well-informed population is more likely to respond to such policies, while – a crucial point – they also tend to have fewer children, reducing the need for heat, cars and carbon at source.

Another of its recommendations is to use nature herself to repair the damage done.

The United Kingdom has half the average area of forest found in the rest of Europe, which is a much greener continent than a century ago. Belatedly, Britain has joined in, with ambitious planting programmes. A 'national forest' was established in 1995 and has begun to populate parts of the Midlands with a carefully planned photosynthetic factory that will extract carbon from the air, protect the soil, and might even cheer up a few of the inhabitants of Burton-on-Trent. In 2018 emerged a grander scheme for an array of fifty million broad-leaved trees that will stretch from the Irish Sea to the North Sea, encompassing Liverpool,

Manchester and Leeds as it does so. Scotland, too, hopes to increase cover by half, although it faces complaints from the owners of grouse moors and deer forests that this interferes with their rights of slaughter.

Such planted woodlands are less diverse, and trap less carbon, than are those that regenerate naturally. The British landscape has been formed by man, and almost none of the 'wildwood' that once covered much of it is left.

My schooldays were – although nobody ever told us – a reminder of what we have lost. In the fourteenth-century poem *Sir Gawain and the Green Knight*, the eponymous hero takes his life in his hands with a journey across the Wirral peninsula, described in the poem as the 'wyldreness of Wyrale', a place sparsely populated with wild men called woodwose. He has an appointment to have his head struck off, and encounters wolves and worse in the forest that cloaked it on his way to fulfil it (there is a twist in the tale that saves his bacon). Now the forest has been replaced by golf-courses.

On Britain's uplands even more damage has been done to the landscape. Many of the hills have been stripped bare by sheep, to form true wildernesses.

There have been several outbreaks of disease in recent years and the sheep have been removed. Within a few months, the landscape begins to regenerate, with flowers and saplings. By simply doing nothing except keeping the woolly creatures out, in a couple of decades there would emerge a thick scrub that would in time evolve into a dense and diverse forest that sucks in carbon dioxide, and by retaining water, reduces flooding.

So far, thanks to the machinations of the land-owners, there is no sign that this will be allowed to happen.

Elsewhere, grander schemes are under way. Since the millennium, China has made an annual investment of fifteen billion dollars in reforestation. Natural forests are protected from loggers, and thousands of new trees are planted each year, on an area the size of Ireland. The authorities hope to increase the area of woodland by forty million hectares within the next few years. The planners have put a tax on disposable chopsticks to draw attention to the issue.

Another approach recommended by the IPCC is to use carbon capture and to store the resulting greenhouse gases underground. The technique has long been used to flush oil from underground reserves, but carbon dioxide is now being hidden under the sea in subterranean salt beds. In the past four decades, some two hundred megatons (a tiny fraction of the total output) has been buried. The gas is extracted from smoke or even from the air itself using filters that trap and bind it, and release it when heated up. More than forty such projects are planned or under way, but to meet the demands of the IPCC fifty times as many will be needed.

A related approach is to use chemistry to transform natural gas – methane – mixed with steam to make a mixture of carbon dioxide and hydrogen. The carbon dioxide is then extracted and pumped underground while the hydrogen is used as fuel.

Biology can also do its bit. An Illinois unit that makes bio-ethanol – a 'green' supplement to gasoline – takes maize and ferments it. The emissions from its vats are pumped into porous sandstone, which will store it for centuries. It has injected a million tons of the stuff, but, again, that is a tiny proportion of what is needed.

Various other proposals are still under discussion. Geo-engineers hope to manipulate the physics of the Earth in an attempt to save the climate. Their ideas range from the (just) feasible, to the difficult, to the almost impossible.

One idea is to change its albedo, in the hope that a shiny planet will have a brighter future. Agriculture has already done a lot, for farmland absorbs much less solar radiation than did the wild landscapes it replaced. In some places, the farmers have gone further. Almeria, on the southern coast of Spain, has since the 1980s become the Costa del Plàstico, a vast sea of white polythene. Its greenhouses feed Europe but have also changed the weather.

Viewed from space, Almeria stands out as a bright star on the shores of the Mediterranean. The plastic tide has pushed up its reflectivity by around a tenth. Local air temperature has fallen by a degree, while around it the mercury has continued its rise. In China, an area ten times that of Wales is now covered in plastic film, either as greenhouses or as soil mulch. Ecologists and aesthetes may complain, but that pushes up the albedo by as much as a third and may in time slow the rate of climate change.

The villages of southern Spain have white walls and roofs that reflect the sun. Half a million square metres of New York roofs have now, in the same tradition, been shielded by white rubber membranes. One part in eight of Los Angeles is black asphalt, much of it in parking lots. Some have been painted white in an attempt to make its summers less insufferable.

A riskier option plans to increase the reflectivity not of the ground, but of the air. One scheme is to spray seawater into the sky. As it evaporates, water molecules will settle around salt crystals to form a cloud. An even more radical plan is to shoot billions of particles of sulphur compounds into the stratosphere. Optimists claim that to insert a volcano's worth of the element every year for a century or so would halve the expected rise in temperature. However, the sulphur might interfere with the ozone layer and a new era of acid rain might not be welcomed by conservationists (or asthmatics).

Other schemes call for the addition of iron salts to areas of the tropical ocean where that element, essential for plant growth, is in short supply in the hope that this will generate an algal bloom to pull in carbon, or to use vast quantities of limestone and other minerals to reduce acidity and improve its ability to soak up carbon dioxide.

Such grand plans have had plenty of publicity, but in practice have done very little. At least for the present, and perhaps for the foreseeable future, the response to climate change has largely been one of words rather than deeds, of change indefinitely delayed, and, for some states, a paralysing sense of imagined oppression by external enemies.

Such a response is a classic of what ecologists call 'the tragedy of the commons', the observation that grazing land shared by a community is often damaged, and sometimes destroyed, by over-exploitation. Each herder puts his own interests first, whatever the costs to the group as a whole. That issue has resonated throughout history, and appears repeatedly in this book.

The dire state of the Ganges and of the aquifers of the world, the loss of the fertile soil in the interests of a few seasons' crops, the South American droughts that come from the destruction of the Amazon rainforest, the insistence on beefsteak and tuna rather than on chicken or sardines, the extinction of Steller's sea cow to provide Russians with hats, the refusal to accept the existence of rickets among the poor because the rich did not experience it, the insistence on shift work whatever the effects on the health of those obliged to carry it out – all turn on the endless question as to how best to share the threats to our own planet and to ourselves that come from human activity. So far there has been only moderate progress.

It may not be a coincidence that one of the key tactics of the coal and oil lobbyists has been to confuse the evidence on climate change and question the honesty of those who study it: doubt is, indeed, still their product.

One ploy used by that group is to label themselves not as 'deniers' but as 'sceptics' about man-made climate change, overfishing, the damage done by junk food, the resurgence of rickets, the destruction of rainforests and the coming crisis in the water supply. The difference between the two terms might seem small, but is in truth fundamental. All scientists are by their nature pessimistic about what will happen to their research for, more often than not, it is ignored, forgotten by their intended audience, found to be wanting results, or almost impossible to explain.

In homage to that central truth, I cannot, in the interests of balance, real rather than false, finish this book without a mention of a piece of information that emerges from my own work and may clash with the consensus about the effects of climate change. It also hints at the difficulties of making prophecies about living systems as they face a changing world.

Botanists have long been keen on mountains, and a 2016 study revisited three hundred peaks in Europe from Svalbard through Scandinavia and Scotland to the Carpathians, the Alps and the Pyrenees. All those ranges had been surveyed in detail for a century and more. On nine out of every ten, the number of plant species found at the summit had increased as lowland forms climbed upwards to follow the shift in the climate and to add weight to the arguments of those who warn about the ecological dangers we all face.

Other creatures have been rather more capricious in their adherence to the accepted view. Forty years ago, as the idea of global warming began to emerge, we began to ask what had happened to the genes for shell pattern in our *Cepaea* snails over the decades, given their response to changes in temperature on scales from centimetres to kilometres.

The chalk uplands of the Marlborough Downs in Wiltshire are somewhat of a Mecca for students of molluscs, and the distribution of genes in more than a hundred populations of my own favourite species were mapped out in the 1960s.

There have been many changes since then. Places once

grazed are now ploughed, and what was short grass has been replaced by scrub. Trees have been lost, and gallops have turned natural vegetation into suburban lawns (their proprietors are, we found, intensely suspicious about anyone who gets too close to their racehorses). In addition, fifty years ago there were thousands of broken shells scattered around stones used as thrush anvils. Now, bird numbers have crashed and the anvils have almost gone.

In the 1980s, the 1990s, and for (probably) the last time in 2012 we resampled all those populations to see if there had been shifts in the frequencies of the genes involved, perhaps either in response to climate or to other ecological changes. There were none, and a wider check across Europe of hundreds of modern versus older samples showed that, with the exception of a few random local fluctuations, the same was true there.

So far, so unhelpful, but that period represents no more than twenty or so generations of the animals (the equivalent of a one-year laboratory experiment on fruit-flies), so that perhaps there has not yet been time for the genes to respond.

Somewhat to our dismay, there then emerged an unexpected twist in the tale. *Cepaea nemoralis* has a close relative called *Cepaea hortensis*, which shares many of its shell patterns. *Hortensis* is a northern species, found almost to the Arctic Circle, while *nemoralis* gives up at around the latitude of Aberdeen. In the south, *hortensis* reaches no further than the Pyrenees, while *nemoralis* is found deeper into Spain. On the Downs in the 1960s *hortensis* was the junior partner of its better-known neighbour, for it was restricted to the lower slopes of the area's hills, perhaps because, as in the basins of Croatia's Velebit Mountains, cool air accumulates there at night.

In our surveys of genetic changes in *Cepaea nemoralis* over half a century, we found – to our surprise, and in spite of an indubitable rise in temperature – that the northern species has shown an inexorable advance towards the modest summits of the Downs and has replaced its southern relative over much of the area. If it continues to spread it will drive *nemoralis* from its ancestral homeland on the Downs and from many other places within the next few decades. When it comes to the reason why, I find myself baffled. Perhaps changes in land use are involved, or perhaps we have got it wrong about the relative thermal tolerance of the two species, but we do not know.

Those sceptical about the effects of man-made global warming may wish to seize on this apparent anomaly to throw doubt on the idea. I prefer to use my favourite and much-quoted phrase of Darwin, that 'ignorance more frequently breeds confidence than does knowledge' as a rallying call against those who deny the truth.

If nothing else, that observation shows that biology and uncertainty are bedfellows and that quite often the more one learns about ecology or molecular genetics the less one seems to understand. That is also true for many of the interactions of the world of life with our closest star. Even so, it has become hard to deny that today's shifts in the thermometer have come at least in part from human interference and that they have had profound effects on the world of life. Given the role of the sun's rays in health, in happiness, in memory, in food, in water, in the shape of the world around us, and in our very existence, its behaviour in the next few decades should become a topic of almost as much interest to men and women as it has long been to snails.

ACKNOWLEDGEMENTS

The scientific work described in this book and in several places referred to as my own could never have been done without the help of large numbers of friends and colleagues. They include Robert Cowie, Jerry Coyne, Tessa Day, Tom Day, David Ellis, Chris Jackson, Dick Lewontin, Leslie Noble, Howard Ochman, Steve Oldfield, Linda Partridge, Tim Prout, Robert Selander, Chuck Taylor and Michael Turelli, together with a host of friends and students who have helped in the field and in the laboratory. My tutor and PhD supervisor, the late Bryan Clarke, was an inspiration for me and for many others, with a readiness to engage with students in a manner impossible today and an infectious enthusiasm which inspired a generation of young geneticists. *Here Comes the Sun* is in part an attempt to give him the thanks I failed sufficiently to express when he was still with us.

INDEX